Statistical Thinking

Statistical Thinking

ANALYZING DATA IN AN UNCERTAIN WORLD

RUSSELL A. POLDRACK

PRINCETON UNIVERSITY PRESS

PRINCETON & OXFORD

Published by Princeton University Press
41 William Street, Princeton, New Jersey 08540
99 Banbury Road, Oxford OX2 6JX

press.princeton.edu

Library of Congress Cataloging-in-Publication Data

Names: Poldrack, Russell A., author.
Title: Statistical thinking : analyzing data in an uncertain world / Russell A. Poldrack.
Description: Princeton : Princeton University Press, [2023] |
    Includes bibliographical references and index.
Identifiers: LCCN 2022053143 (print) | LCCN 2022053144 (ebook) | ISBN 9780691250939
    (hardback; alk. paper) | ISBN 9780691218441 (paperback; alk. paper) |
    ISBN 9780691230825 (ebook)
Subjects: LCSH: Statistics. | Social sciences—Statistical methods.
Classification: LCC HA29 .P634 2023 (print) | LCC HA29 (ebook) |
    DDC 519.5—dc23/eng/20221104
LC record available at https://lccn.loc.gov/2022053143
LC ebook record available at https://lccn.loc.gov/2022053144

British Library Cataloging-in-Publication Data is available

Editorial: Hallie Stebbins and Kiran Pandey
Production Editorial: Natalie Baan
Cover Design: Wanda España
Production: Danielle Amatucci
Publicity: William Pagdatoon
Copyeditor: Jennifer McClain

Cover image: Alan Rogerson / Alamy Stock Photo

This book has been composed in Arno and sans

10 9 8 7 6 5 4 3 2 1

CONTENTS

PREFACE

The goal of this book is to the tell the story of statistics as it is used today by researchers around the world. It's a different story than the one told in most introductory statistics books, which focus on teaching how to use a set of tools to achieve very specific goals. This book focuses on understanding the basic ideas of *statistical thinking*—a systematic way of thinking about how we describe the world and use data to make decisions and predictions, all in the context of the inherent uncertainty that exists in the real world. It also brings to bear current methods that have only become feasible in light of the amazing increases in computational power that have happened in the last few decades. Analyses that would have taken years in the 1950s can now be completed in a few seconds on a standard laptop computer, and this power unleashes the ability to use computer simulation to ask questions in new and powerful ways.

The book is also written in the wake of the reproducibility crisis that has engulfed many areas of science since 2010. One of the important roots of this crisis is found in the way that statistical hypothesis testing has been used by researchers (and sometimes abused, as I detail in the final chapter of the book), and this ties directly to statistical education. Thus, a goal of the book is to highlight the ways in which some current statistical methods may be problematic, and to suggest alternatives.

The Golden Age of Data

Throughout this book I have tried when possible to use examples from real data. This is now very easy because we are swimming in open datasets, as governments, scientists, and companies are increasingly making data freely available. Using real datasets is important because it prepares students to work with real data, which I think should be one of the major goals of statistical training. It also helps us realize (as we will see at various points throughout the book) that data don't always come to us ready to analyze, and often need *wrangling* to help get them into shape. Using real data also shows that the idealized statistical distributions often assumed in statistical methods don't always hold in the real world—for example, as we will see in chapter 3, distributions of some real-world quantities (like the number of friends on Facebook) can have very long tails that can break many standard assumptions.

I apologize up front that the datasets are heavily US-centric, which is due in large part to the substantial amount of data made freely available by US government agencies. We rely heavily on the very rich National Health and Nutrition Examination Surveys (NHANES) dataset that is available as an R package, as well as a number of complex datasets included in R (such as those in the `fivethirtyeight` package) that are also based in the US. I hope other countries will begin to share data more freely so that future editions can use a more diverse range of datasets.

The Importance of Doing Statistics

The only way to really learn statistics is to *do* statistics. While many statistics courses historically have been taught using point-and-click statistical software, it is increasingly common for statistical education to use open source languages in which students can code their own analyses. I think that being able to code one's analyses is essential in order to gain a deep appreciation for statistical analysis, which is why the students in my course at Stanford are expected to learn to use the R statistical programming language to analyze data, alongside the theoretical knowledge that they learn from this book.

There are two openly available companions to this textbook that can help the reader get started learning to program; one focuses on the R programming language, and another focuses on the Python language. These can be accessed via https://press.princeton.edu /isbn/9780691218441, under Resources, along with examples of how to create the book's figures in both Python and R.

Acknowledgments

I'd like to start by thanking Susan Holmes, who first inspired me to consider writing my own statistics book. I would also like to thank Jeanette Mumford, who provided very helpful suggestions on the entire book and substantial help in developing the content for several sections. Anna Khazenzon provided early comments and inspiration. Lucy King provided detailed comments and edits on the entire book, and helped clean up the code so that it was consistent with the Tidyverse. Michael Henry Tessler provided very helpful comments on the Bayesian analysis chapter. Particular thanks also go to Yihui Xie, creator of the Bookdown package, for helping in the use of Bookdown features to create the book.

I'd also like to thank others who provided helpful comments and suggestions: Athanassios Protopapas, Wesley Tansey, Jack Van Horn, Thor Aspelund, and Jessica Riskin, as well as @enoriverbend.

This book grew out of an open source version that remains freely available online at http://statsthinking21.org. Thanks to the following individuals who have contributed edits or issues to the open source version of the book by Github or email: Isis Anderson, Larissa Bersh, Isil Bilgin, Forrest Dollins, Chuanji Gao, Nate Guimond, Alan He, Wu Jianxiao, James Kent, Dan Kessler, Philipp Kuhnke, Leila Madeleine, Lee Matos, Ryan McCormick, Jarod Meng, Kirsten Mettler, Shanaathanan Modchalingam, Mehdi Rahim,

Jassary Rico-Herrera, Martijn Stegeman, Mingquian Tan, Wenjin Tao, Laura Tobar, Albane Valenzuela, Alexander Wang, and Michael Waskom, as well as barbyh, basicv8vc, brettelizabeth, codetrainee, dzonimn, epetsen, carlosivanr, hktang, jiamingkong, khtan, kiyofumi-kan, NevenaK, and ttaweel.

Particular thanks go to Isil Bilgin for assistance in fixing many of these issues on the open source version of the book.

Statistical Thinking

1

Introduction

Learning Objectives

Having read this chapter, you should be able to

- Describe the central goals and fundamental concepts of statistics.
- Describe the difference between experimental and observational research with regard to what can be inferred about causality.
- Explain how randomization provides the ability to make inferences about causation.

What Is Statistical Thinking?

Statistical thinking is a way of understanding a complex world by describing it in relatively simple terms that nonetheless capture essential aspects of its structure or function, and that also provide us with some idea of how uncertain we are about that knowledge. The foundations of statistical thinking come primarily from mathematics and statistics but also from computer science, psychology, and other fields of study.

We can distinguish statistical thinking from other forms of thinking that are less likely to describe the world accurately. In particular, human intuition frequently tries to answer the same questions that we can answer using statistical thinking, but it often gets the answer wrong. For example, in recent years most Americans have reported that they think violent crime is worse in the current year compared to the previous year (Pew 2020). However, a statistical analysis of the actual crime data showed that in fact violent crime was steadily *decreasing* during that time. Intuition fails us because we rely on best guesses (which psychologists refer to as *heuristics*) that can often get it wrong. For example, humans often judge the prevalence of some event (like violent crime) using an *availability heuristic*—that is, how easily we can think of an example of violent crime. For this reason, our judgments of increasing crime rates may be more reflective of increasing news coverage, in spite of an actual decrease in the rate of crime. Statistical thinking provides us with the tools to more accurately understand the world and overcome the biases of human judgment.

Dealing with Statistics Anxiety

Many people come to their first statistics class with a lot of trepidation and anxiety, especially once they hear that they will also have to learn to code in order to analyze data. In my class, I give students a survey prior to the first session to measure their attitude toward statistics, asking them to rate a number of statments on a scale of 1 (strongly disagree) to 7 (strongly agree). One of the items on the survey is "The thought of being enrolled in a statistics course makes me nervous." In a recent class, almost two-thirds of the students responded with a 5 or higher, and about one-fourth of the students said they strongly agreed with the statement. So if you feel nervous about starting to learn statistics, you are not alone.

Anxiety feels uncomfortable, but psychology tells us that this kind of emotional arousal can actually help us perform *better* on many tasks, by focusing our attention. So if you start to feel anxious about the material in this book, remind yourself that many other readers are feeling similarly, and that this emotional arousal could actually help cement the material in your brain more effectively (even if it doesn't seem like it!).

What Can Statistics Do for Us?

There are three major things that we can do with statistics:

- *Describe*: The world is complex and we often need to describe it in a simplified way that we can understand.
- *Decide*: We often need to make decisions based on data, usually in the face of uncertainty.
- *Predict*: We often wish to make predictions about new situations based on our knowledge of previous situations.

Let's look at an example of these in action, centered on a question that many of us are interested in: How do we decide what's healthy to eat? There are many different sources of guidance: government dietary guidelines, diet books, and bloggers, just to name a few. Let's focus in on a specific question: Is saturated fat in our diet a bad thing?

One way that we might answer this question is common sense. If we eat fat, then it's going to turn straight into fat in our bodies, right? And we have all seen photos of arteries clogged with fat, so eating fat is going to clog our arteries, right?

Another way that we might answer this question is by listening to authority figures. The dietary guidelines from the US Food and Drug Administration have as one of their key recommendations that "A healthy eating pattern limits saturated fats." You might hope that these guidelines would be based on good science, and in some cases they are; but as Nina Teicholz (2014) outlined in her book *Big Fat Surprise*, this particular recommendation seems to be based more on the long-standing dogma of nutrition researchers than on actual evidence.

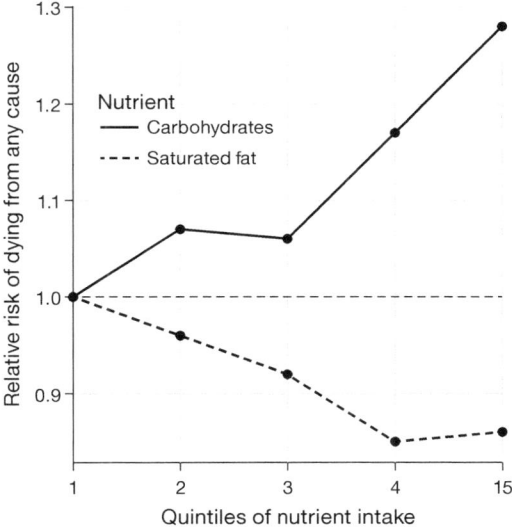

FIGURE 1.1. A plot of data from the PURE study, showing the relationship between death from any cause and the relative intake of saturated fats and carbohydrates.

Finally, we might look at actual scientific research. Let's start by looking at the PURE study, which has examined diets and health outcomes (including death) in more than 135,000 people from 18 different countries. In one of the analyses of this dataset (Dehghan et al. 2017, published in *the Lancet*), the PURE investigators reported an analysis of how intake of various classes of macronutrients (including saturated fats and carbohydrates) was related to the likelihood of people dying during the time they participated in the study. Participants were followed for a *median* of 7.4 years, meaning that half of the people in the study were followed for less than 7.4 years and half were followed for more than 7.4 years. Figure 1.1 plots some of the data from the study (extracted from the publication), showing the relationship between the intake of both saturated fats and carbohydrates and the risk of dying from any cause.

This plot is based on 10 numbers. To obtain these numbers, the researchers split the group of 135,335 study participants (which we call the *sample*) into five groups (quintiles) after ordering them in terms of their intake of either of the nutrients; the first quintile contains the 20% of people with the lowest intake, and the fifth quintile contains the 20% with the highest intake. The researchers then computed how often people in each of those groups died during the time they were being followed. The figure expresses this in terms of the *relative risk* of dying in comparison to the lowest quintile: if this number is greater than one, it means that people in the group are *more* likely to die than are people in the lowest quintile, whereas if it's less than one, it means that people in the group are *less* likely to die. The figure is pretty clear: people who ate more saturated fat were *less* likely to die during the study, with the lowest death rate seen for people who were in the fourth quintile (that is, those who ate more fat than the lowest 60% but less than the top 20%). The opposite is seen for carbohydrates; the more carbs a person ate, the more likely they were to die during the study. This example shows how we can use statistics to *describe* a complex dataset in terms of a much simpler set of numbers; if we had to look at the data from each of the study participants at the same time, we would be overloaded with data and it would be hard to see the pattern that emerges when they are described more simply.

The numbers in figure 1.1 seem to show that deaths decrease with saturated fat and increase with carbohydrate intake, but we also know that there is a lot of uncertainty in the data; there are some people who died early even though they ate a low-carb diet, and, similarly, some people who ate a ton of carbs but lived to a ripe old age. Given this variability, we want to *decide* whether the relationships we see in the data are large enough that we wouldn't expect them to occur randomly if there was not truly a relationship between diet and longevity. Statistics provide us with the tools to make these kinds of decisions, and often people from the outside view this as *the* main purpose of statistics. But as we will see throughout the book, this need for black-and-white decisions based on fuzzy evidence has often led researchers astray.

Based on the data, we would also like to be able to *predict* future outcomes. For example, a life insurance company might want to use data about a particular person's intake of fat and carbohydrates to predict how long they are likely to live. An important aspect of prediction is that it requires us to generalize from the data we already have to some other situation, often in the future; if our conclusions were limited to the specific people in the study at a particular time, then the study would not be very useful. In general, researchers must assume that their particular sample is representative of a larger *population*, which requires that they obtain the sample in a way that provides an unbiased picture of the population. For example, if the PURE study had recruited all of its participants from religious sects that practice vegetarianism, then we probably wouldn't want to generalize the results to people who follow different dietary standards.

The Big Ideas of Statistics

There are a number of very basic ideas that cut through nearly all aspects of statistical thinking. Several of these are outlined by Stigler (2016) in his outstanding book *The Seven Pillars of Statistical Wisdom*, which I have augmented here.

Learning from Data

One way to think of statistics is as a set of tools that enable us to learn from data. In any situation, we start with a set of ideas or *hypotheses* about what might be the case. In the PURE study, the researchers may have started out with the expectation that eating more fat would lead to higher death rates, given the prevailing negative dogma about saturated fats. Later in the book we introduce the idea of *prior knowledge*, that is, the knowledge that we bring to a situation. This prior knowledge can vary in its strength, often based on our level of experience; if I visit a restaurant for the first time, I am likely to have a weak expectation of how good it will be, but if I visit a restaurant where I have eaten 10 times before, my expectations will be much stronger. Similarly, if I look at a restaurant review site and see that a restaurant's average rating of four stars is based on only three reviews, I will have a weaker expectation than I would if it were based on three hundred reviews.

Statistics provides us with a way to describe how new data can be best used to update our beliefs, and in this way there are deep links between statistics and psychology. In fact, many theories of human and animal learning from psychology are closely aligned with ideas from the field of *machine learning*—a new field at the interface of statistics and computer science that focuses on how to build computer algorithms that can learn from experience. While statistics and machine learning often try to solve the same problems, researchers from these fields frequently take very different approaches; the famous statistician Leo Breiman (2001) once referred to them as "the two cultures" to reflect how different their approaches can be. In this book, I try to blend the two cultures together because both approaches provide useful tools for thinking about data.

Aggregation

Another way to think of statistics is as "the science of throwing away data." In the example of the PURE study above, we took more than 100,000 numbers and condensed them into 10. It is this kind of *aggregation* that is one of the most important concepts in statistics. When it was first advanced, this idea was revolutionary: if we throw out all the details about every one of the participants, then how can we be sure we aren't missing something important?

As we will see, statistics provides us with ways to characterize the structure of aggregations of data, with theoretical foundations that explain why this usually works well. However, it's also important to keep in mind that aggregation can go too far, and later we will encounter cases where a summary can provide a very misleading picture of the data being summarized.

Uncertainty

The world is an uncertain place. We now know that cigarette smoking causes lung cancer, but this causation is probabilistic: a 68-year-old man who smoked two packs a day for the past 50 years and continues to smoke has a 15% (1 out of 7) risk of getting lung cancer, which is much higher than the chance of lung cancer in a nonsmoker. However, it also means that there will be many people who smoke their entire lives and never get lung cancer. Statistics provides us with the tools to characterize uncertainty, to make decisions under uncertainty, and to make predictions whose uncertainty we can quantify.

One often sees journalists write that scientific researchers have "proved" some hypothesis. But statistical analysis can never "prove" a hypothesis, in the sense of demonstrating that it must be true (as one would in a logical or mathematical proof). Statistics can provide us with evidence, but it's always tentative and subject to the uncertainty that is ever present in the real world.

Sampling from a Population

The concept of aggregation implies that we can make useful insights by collapsing across data—but how much data do we need? The idea of *sampling* says that we can summarize an entire population based on just a small number of samples from the population, as long as those samples are obtained in the right way. For example, the PURE study enrolled a sample of about 135,000 people, but its goal was to provide insights about the billions of humans who make up the population from which those people were sampled. As we already discussed above, the way that the study sample is obtained is critical, as it determines how broadly we can generalize the results. Another fundamental insight about sampling is that, while larger samples are always better (in terms of their ability to accurately represent the entire population), there are diminishing returns as the sample gets larger. In fact, the rate at which the benefit of larger samples decreases follows a simple mathematical rule, growing as the square root of the sample size, such that in order to double the precision of our estimate we need to quadruple the size of our sample.

Causality and Statistics

The PURE study seemed to provide pretty strong evidence for a positive relationship between eating saturated fat and living longer, but this doesn't tell us what we really want to know: If we eat more saturated fat, will that cause us to live longer? This is because we don't know whether there is a direct causal relationship between eating saturated fat and living longer. The data are consistent with such a relationship, but they are equally consistent with some other factor causing both higher saturated fat and longer life. For example, one might imagine that people who are richer eat more saturated fat and richer people tend to live longer, but their longer life is not necessarily due to fat intake—it could instead be due to better health care, reduced psychological stress, better food quality, or many other factors. The PURE study investigators tried to account for these factors, but we can't be certain that their efforts completely removed the effects of other variables. The fact that other factors may explain the relationship between saturated fat intake and death is an example of why introductory statistics classes often teach that "correlation does not imply causation," though the renowned data visualization expert Edward Tufte has added, "but it sure is a hint."

Although observational research (like the PURE study) cannot conclusively demonstrate causal relationships, we generally think that causation can be demonstrated using studies that experimentally control and manipulate a specific factor. In medicine, such a study is referred to as a *randomized controlled trial* (RCT). Let's say that we wanted to do an

RCT to examine whether increasing saturated fat intake increases life span. To do this, we would sample a group of people and then assign them to either a treatment group (which would be told to increase their saturated fat intake) or a control group (who would be told to keep eating the same as before). It is essential that we assign the individuals to these groups randomly. Otherwise, people who choose the treatment group might be different in some way than people who choose the control group—for example, they might be more likely to engage in other healthy behaviors as well. We would then follow the participants over time and see how many people in each group died. Because we randomized the participants to treatment or control groups, we can be reasonably confident that there are no other differences between the groups that would *confound* the treatment effect; however, we still can't be certain because sometimes randomization yields treatment groups versus control groups that *do* vary in some important way. Researchers often try to address these confounds using statistical analyses, but removing the influence of a confound from the data can be very difficult.

A number of RCTs have examined the question of whether changing saturated fat intake results in better health and longer life. These trials have focused on *reducing* saturated fat because of the strong dogma among nutrition researchers that saturated fat is deadly; most of these researchers would have probably argued that it was not ethical to cause people to eat *more* saturated fat! However, the RCTs have shown a very consistent pattern: overall there is no appreciable effect on death rates of reducing saturated fat intake.

Suggested Reading

- *The Seven Pillars of Statistical Wisdom*, by Stephen Stigler. Stigler is one of the world's leading historians of statistics, and this book outlines what he sees as a number of the foundataional ideas that underlie statistical thinking. His ideas strongly influenced the basic ideas that are presented in this chapter.
- *The Lady Tasting Tea: How Statistics Revolutionized Science in the Twentieth Century*, by David Salsburg. This book provides a readable yet detailed overview of the history of statistics, with a strong focus on amplifying the often overlooked contributions of women in the history of statistics.
- *Naked Statistics: Stripping the Dread from the Data*, by Charles Wheelan. A very fun tour of the main ideas of statistics.

Problems

1. Describe how statistics can be thought of as a set of tools for learning from data.
2. What does it mean to sample from a population, and why is this useful?
3. Describe the concept of a *randomized controlled trial* and outline the reason that we think that such an experiment can provide information about the causal effect of a treatment.
4. The three things that statistics can do for us are to _____, _____, and _____.

5. Early in the COVID-19 pandemic there was an observational study that reported effectiveness of the drug hydroxycholoroquine in treating the disease. Subsequent randomized controlled trials showed no effectiveness of the drug for treating the disease. What might explain this discrepancy between observational results and randomized controlled trials? Choose all that apply.
   - There were systematic differences in the observational study between those prescribed the drug and those who were prescribed other treatments.
   - Randomization helps eliminate differences between the treatment and control groups.
   - The drug is more effective when the physician gives it to the right patients.

6. Match the following examples with the most appropriate concept from the following list: aggregation, uncertainty, sampling.
   - A researcher summarizes the scores of 10,000 people in a set of 12 numbers.
   - A person smokes heavily for 70 years but remains perfectly healthy and fit.
   - A researcher generalizes from a study of 1000 individuals to all humans.

2

Working with Data

Learning Objectives

Having read this chapter, you should be able to

- Distinguish between different types of variables (quantitative/qualitative, binary/integer/real, discrete/continuous) and give examples of each of these kinds of variables.
- Distinguish between the concepts of reliability and validity and apply each concept to a particular dataset.

What Are Data?

The first important point about data is that data *are*—meaning that the word *data* is the plural form of the singular *datum* (though some people disagree with me on this). You might also wonder how to pronounce *data*—I say "day-tah," but I know many people who say "dah-tah," and I have been able to remain friends with them in spite of this. Now, if I heard them say "the data is," that would be a bigger issue . . . .

Qualitative Data

Data are composed of *variables*, where a variable reflects a unique measurement or quantity. Some variables are *qualitative*, meaning that they describe a quality rather than a numeric quantity. For example, one of the questions that I ask in the introductory survey for my statistics course is, "What is your favorite food?," to which some of the answers have been blueberries, chocolate, tamales, pasta, pizza, and mangoes. Those data are not intrinsically numerical; we could assign numbers to each one ($1 =$ blueberries, $2 =$ chocolate, etc.), but we would just be using the numbers as labels rather than as real numbers. This also constrains what we should do with those numbers; for example, it wouldn't make sense to compute the average of those numbers. However, we often code qualitative data using numbers in order to make them easier to work with, as you will see later.

Table 2.1. Counts of the Prevalence of Different Responses to the Question "Why are you Taking this Class?"

Why are you taking this class?	Number of students
It fulfills a degree plan requirement	105
It fulfills a general education breadth requirement	32
It is not required but I am interested in the topic	11
Other	4

Quantitative Data

More commonly in statistics we work with *quantitative* data, meaning data that are numerical. For example, table 2.1 shows the results from another question that I ask in my introductory class, which is "Why are you taking this class?" Note that the students' answers were qualitative, but we generated a quantitative summary of them by counting how many students gave each response.

TYPES OF NUMBERS

There are several different types of numbers that we work with in statistics. It's important to understand these differences, in part because statistical analysis languages (such as R) often require distinguishing between them.

Binary numbers. The simplest are binary numbers—that is, zero or one. We often use binary numbers to represent whether something is true or false, or present or absent. For example, I might ask 10 people if they have ever experienced a migraine headache, recording their answers as "Yes" or "No." It's often useful to instead use *logical* values, which take the value of either True or False. This can be especially useful when we start using programming languages like R to analyze our data, since these languages already understand the concepts of True and False. In fact, most programming languages treat truth values and binary numbers equivalently. The number one is equal to the logical value TRUE, and the number zero is equal to the logical value FALSE.

Integers. Integers are whole numbers with no fractional or decimal part. We most commonly encounter integers when we count things, but they also often occur in psychological measurement. For example, in my introductory survey I administer a set of questions about attitudes toward statistics (such as "Statistics seems very mysterious to me"), on which the students respond with a number between 1 (disagree strongly) and 7 (agree strongly).

Real numbers. Most commonly in statistics we work with real numbers, which have a fractional/decimal part. For example, we might measure someone's weight, which can be measured to an arbitrary level of precision, from kilograms down to micrograms.

Discrete versus Continuous Measurements

A *discrete* measurement is one that takes one of a finite set of particular values. These could be qualitative values (for example, different breeds of dogs) or numerical values (for example, how many friends one has on Facebook). Importantly, there is no middle ground between the measurements; it doesn't make sense to say that one has 33.7 friends.

A *continuous* measurement is one that is defined in terms of a real number. It could fall anywhere in a particular range of values, though usually our measurement tools limit the precision with which we can measure it; for example, a floor scale might measure weight to the nearest pound, even though weight could in theory be measured with much more precision.

It is common in statistics courses to go into more detail about different "scales" of measurement, which are discussed in the appendix to this chapter. The most important takeaway is that some kinds of statistics don't make sense on some kinds of data. For example, imagine that we were to collect postal zip code data from a number of individuals. Those numbers are represented as integers, but they don't actually refer to a numeric scale; each zip code basically serves as a label for a different region. For this reason, it wouldn't make sense to talk about the average zip code, for example.

What Makes a Good Measurement?

In many fields, such as psychology, the thing we are measuring is not a physical feature but instead is an unobservable theoretical concept, which we usually refer to as a *construct*. For example, let's say that I want to test how well an individual understands the distinction between the different types of numbers described above. I could give them a pop quiz that would ask several questions about these concepts and count how many they got right. This test might or might not be a good measurement of the construct of their actual knowledge—for example, if I were to write the test in a confusing way or use language that they don't understand, then the test might suggest they don't understand the concepts when really they do. On the other hand, if I give a multiple choice test with very obvious wrong answers, they might be able to perform well on the test even if they don't actually understand the material.

It is usually impossible to measure a construct without some amount of error. In the example above, you might know the answer, but you might misread the question and get it wrong. In other cases, there is error intrinsic to the thing being measured, such as when we measure how long it takes a person to respond on a simple reaction time test, which will vary from trial to trial for many reasons. We generally want our measurement error to be as low as possible, which we can acheive either by improving the quality of the measurement (for example, using a more accurate device to measure reaction time) or by averaging over a larger number of individiual measurements.

Sometimes there is a standard against which other measurements can be tested, which we might refer to as a "gold standard"—for example, measurement of sleep can be done using many different devices (such as those that measure movement in bed), but they are generally considered inferior to the gold standard of polysomnography (which uses measurement of brain waves to quantify the amount of time a person spends in each stage of sleep). Often the gold standard is more difficult or expensive to perform, and the cheaper method is used even though it might have greater error.

When we think about what makes a good measurement, we usually distinguish two different aspects: it should be *reliable* and it should be *valid*.

Reliability

Reliability refers to the consistency of our measurements. One common form of reliability, known as *test-retest reliability*, measures how well the measurements agree if the same measurement is performed twice. For example, I might give you a questionnaire about your attitude toward statistics today, repeat this same questionnaire tomorrow, and compare your answers on the two days; we would hope they would be very similar to one another, unless something happened in between the two tests that should have changed your view of statistics (like reading this book!).

Another way to assess reliability comes in cases where the data include subjective judgments. For example, let's say that a researcher wants to determine whether a treatment changes how well an autistic child interacts with other children, which is measured by having experts watch the child and rate their interactions with the other children. In this case, we would like to make sure that the answers don't depend on the individual rater—that is, we would like to have high *interrater reliability*. This can be assessed by having more than one rater perform the rating and then comparing the ratings to make sure that they agree well with one another.

Reliability is important if we want to compare one measurement to another, because the relationship between two different variables can't be any stronger than the relationship between either of the variables and itself (i.e., its reliability). This means that an unreliable measure can never have a strong statistical relationship with any other measure. For this reason, researchers developing a new measurement (such as a new survey) often go to great lengths to establish and improve its reliability.

Validity

Reliability is important, but on its own it's not enough: after all, I could create a perfectly reliable measurement on a personality test by recoding every answer using the same number, regardless of how the person actually answers. We want our measurements to also be *valid*—that is, we want to make sure that we are actually measuring the construct we think we are measuring (figure 2.1). There are many different types of validity that are commonly discussed; we focus on three of them.

 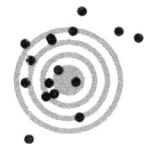

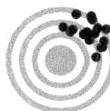 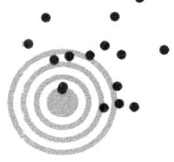

FIGURE 2.1. A figure demonstrating the distinction between reliability and validity, using shots at a bull's-eye. Reliability refers to the consistency of the location of shots, and validity refers to the accuracy of the shots with respect to the center of the bull's-eye.

FACE VALIDITY

Does the measurement make sense on its face? If I were to tell you that I was going to measure a person's blood pressure by looking at the color of their tongue, you would probably think that this was not a valid measure on its face. On the other hand, using a blood pressure cuff would have face validity. This is usually a first reality check before we dive into more complicated aspects of validity.

CONSTRUCT VALIDITY

Is the measurement related to other measurements in an appropriate way? This is often subdivided into two aspects. *Convergent validity* means that the measurement should be closely related to other measures that are thought to reflect the same construct. Let's say that I am interested in measuring how extroverted a person is by using either a questionnaire or an interview. Convergent validity would be demonstrated if both of these different measurements are closely related to one another. On the other hand, measurements thought to reflect different constructs should be unrelated, known as *divergent validity*. If my theory of personality says that extroversion and conscientiousness are two distinct constructs, then I should also see that my measurements of extroversion are *unrelated* to measurements of conscientiousness.

PREDICTIVE VALIDITY

If our measurements are truly valid, then they should also be predictive of other outcomes. For example, let's say that we think the psychological trait of sensation seeking (the desire for new experiences) is related to risk taking in the real world. To test for predictive validity of a measurement of sensation seeking, we would test how well scores on the test predict scores on a different survey that measures real-world risk taking.

Suggested Reading

- *An Introduction to Psychometric Theory with Applications in R*, by William Revelle. This is a free online textbook that covers many aspects of measurement theory and its relation to data analysis, by one of the field's most important researchers.

Problems

1. For each of the following variables, identify whether it is qualitative or quantitative:
   - The team memberships of a set of basketball players
   - The number of elections won by a politician
   - The version number of a car (for example, the BMW 330)
   - The proportion of students who have taken a statistics class
   - The name of the textbook used for statistics classes across the country
   - Number of Twitter followers
   - College major
   - Amount of daily rainfall
   - Presence/absence of disease antibodies
2. For each of the following variables, identify what kind of number it reflects (binary, integer, or real):
   - The number of students in a class
   - The proportion of students who are expected to take a course
   - Whether a particular student has or has not completed an assignment
   - The weight of a fruit fly in ounces
3. Provide an example of a case where a number is used to represent information that is not actually on a numeric scale.
4. For each of the following examples, specify whether it reflects the concept of validity or reliability.
   - Two experts are asked to rate the quality of a piece of art, and their ratings are compared for their similarity.
   - Students are given a test with many questions about a particular topic, presented in random order. Their score on the odd-numbered questions is compared to their score on the even-numbered questions.
   - Students are given two different tests for their knowledge, one using multiple choice questions and one using a hands-on test of practical knowledge, and the scores on the two tests are compared.
   - A wearable device for measuring blood glucose is given to a group of individuals, and the measurements for each individual are compared to their subsequent likelihood of developing diabetes.

Appendix

Scales of Measurement

All variables must take on at least two different possible values (otherwise they would be *constants* rather than variables), but different values of the variable can relate to each other in different ways, which we refer to as *scales of measurement*. There are four ways in which the values of a variable can differ.

- *Identity*: Each value of the variable has a unique meaning.
- *Magnitude*: The values of the variable reflect different magnitudes and have an ordered relationship to one another—that is, some values are larger and some are smaller.
- *Equal intervals*: Units along the scale of measurement are equal to one another. This means, for example, that the difference between 1 and 2 would be equal in its magnitude to the difference between 19 and 20.
- *Absolute zero*: The scale has a true meaningful zero point. For example, for many measurements of physical quantities, such as height or weight, this is the complete absence of the thing being measured.

There are four different scales of measurement that go along with the ways that values of a variable can differ.

- *Nominal scale*: A nominal variable satisfies the criterion of identity, such that each value of the variable represents something different, but the numbers simply serve as qualitative labels as discussed above. For example, we might ask people for their political party affiliation and then code those as numbers: 1 = Republican, 2 = Democrat, 3 = Libertarian, and so on. However, the different numbers do not have any ordered relationship with one another.
- *Ordinal scale*: An ordinal variable satisfies the criteria of identity and magnitude, such that the values can be ordered in terms of their magnitude. For example, we might ask a person with chronic pain to complete a form every day assessing how bad their pain is, using a 1–7 numeric scale. Note that while the person is presumably feeling more pain on a day when they report a 6 versus a day when they report a 3, it wouldn't make sense to say that their pain is twice as bad on the former versus the latter day; the ordering gives us information about relative magnitude, but the differences between values are not necessarily equal in magnitude.
- *Interval scale*: An interval scale has all the features of an ordinal scale, but in addition the intervals between units on the measurement scale can be treated as equal. A standard example is physical temperature measured in Celsius or Fahrenheit; the physical difference between 10 and 20 degrees is the same as the physical difference between 90 and 100 degrees, but each scale can also take on negative values.

Table 2.2. Different Scales of Measurement Admit Different Types of Numeric Operations

	Equal/not equal	> / <	+/−	Multiply/divide
Nominal	OK			
Ordinal	OK	OK		
Interval	OK	OK	OK	
Ratio	OK	OK	OK	OK

- *Ratio scale*: A ratio scale variable has all four of the features outlined above—identity, magnitude, equal intervals, and absolute zero. The difference between a ratio scale variable and an interval scale variable is that the ratio scale variable has a true zero point. Examples of ratio scale variables include physical height and weight, along with temperature measured in kelvins.

There are two important reasons that we must pay attention to the scale of measurement of a variable. First, the scale determines what kind of mathematical operations we can apply to the data (table 2.2). A nominal variable can only be compared for equality; that is, do two observations on that variable have the same numeric value? It would not make sense to apply other mathematical operations to a nominal variable since they don't really function as numbers in a nominal variable but rather as labels. With ordinal variables, we can also test whether one value is greater or less than another, but we can't do any arithmetic. Interval and ratio variables allow us to perform arithmetic; with interval variables we can only add or subtract values, whereas with ratio variables we can also multiply and divide values.

These constraints also imply that there are certain kinds of statistics that we can compute on each type of variable. Statistics that simply involve counting of different values (such as the most common value, known as the *mode*) can be calculated on any of the variable types. Other statistics are based on ordering or ranking of values (such as the *median*, which is the middle value when all the values are ordered by their magnitude), and these require that the value at least be on an ordinal scale. Finally, statistics that involve adding up values (such as the average, or *mean*) require that the variables be at least on an interval scale. Having said that, we should note that it's quite common for researchers to compute the mean of variables that are only ordinal (such as responses on personality tests), but this can sometimes be problematic.

3

Summarizing Data

I mentioned in chapter 1 that one of the big discoveries of statistics is the idea that we can better understand the world by throwing away information, and that's exactly what we are doing when we summarize a dataset. In this chapter, we discuss why and how to summarize data.

Learning Objectives

Having read this chapter, you should be able to

- Compute absolute, relative, and cumulative frequency distributions for a given dataset.
- Generate a graphical representation of a frequency distribution.
- Describe the difference between a normal and a long-tailed distribution and the situations that commonly give rise to each.

Why Summarize Data?

When we summarize data, we are necessarily throwing away information, and one might plausibly object to this. As an example, let's go back to the PURE study that we discussed in chapter 1. Are we not supposed to believe that all the details about each individual matter, beyond those that are summarized in the dataset? What about the specific details of how the data were collected, such as the time of day or the mood of the participant? All of these details are lost when we summarize the data.

One reason that we summarize data is that it provides us with a way to *generalize*—that is, to make general statements that extend beyond specific observations. The importance of generalization was highlighted by the writer Jorge Luis Borges in his short story "Funes the Memorious," which describes an individual who loses the ability to forget. Borges focuses in on the relationship between generalization (i.e., throwing away data) and thinking: "To think is to forget a difference, to generalize, to abstract. In the overly replete world of Funes, there were nothing but details."

FIGURE 3.1. A Sumerian tablet from the
Louvre Museum in Paris, showing a sales
contract for a house and field. (Public
domain, via Wikimedia Commons)

Psychologists have long studied all the ways in which generalization is central to thinking. One example is categorization: we are able to easily recognize different examples of
the category "birds" even though the individual examples may be very different in their
surface features (such as an ostrich, a robin, and a chicken). Importantly, generalization
lets us make predictions about these individuals—in the case of birds, we can predict that
they can fly and eat seeds and that they probably can't drive a car or speak English. These
predictions won't always be right, but they are often good enough to be useful in the world.

Summarizing Data Using tables

A simple way to summarize data is to generate a table representing counts of various types
of observations. This type of table has been used for thousands of years (figure 3.1).

Let's look at some examples of the use of tables, using a more realistic dataset. Throughout this book we use the National Health and Nutrition Examination Survey (NHANES)
dataset. This is an ongoing study that assesses the health and nutrition status of a sample of individuals from the United States on many different variables. We use a version
of the dataset that is available for the R statistical software package. For this example, we
look at a simple variable, called *PhysActive* in the dataset. This variable contains one of
three different values: "Yes" or "No" (indicating whether or not the person reports engaging in "moderate or vigorous-intensity sports, fitness or recreational activities"), or "NA"
if the data are missing for that individual. There are different reasons that the data might
be missing; for example, this question was not asked of children younger than 12 years of
age, and some adults may have declined to answer the question during the interview, or
the interviewer's recording of the answer on their form might be unreadable.

Table 3.1. Frequency Distribution
for the *PhysActive* Variable

PhysActive	Absolute frequency
No	2473
Yes	2972
NA	1334

When data are missing, it is important to know why. If they are missing for reasons that are not related to the true value of the variable, then we can often simply go forward with analysis of the data, leaving out those missing values. This can occur, for example, if an interviewer forgets to ask a question of some people. In other cases, the fact that the data are missing might actually be related to the value of the variable; for example, if there is a sensitive question about mental health, people with mental health problems might be more likely to skip the question rather than to respond. In those cases, analyzing the data becomes much more challenging, and it's advisable to speak to an expert for further guidance. In many cases, we simply don't know why data are missing; it's common to simply move ahead assuming that they are missing completely at random, but it's important to know that this can cause biases in the results if the data are actually missing in relation to the true value of the variable. For much more on the treatment of missing data, see Enders (2022).

Frequency Distributions

A *distribution* describes how data are distributed between different possible values. For this example, let's look at how many people fall into each of the physical activity categories.

Table 3.1 shows the frequencies of each of the different values; there were 2473 individuals who responded "No" to the question, 2972 who responded "Yes," and 1334 for whom no response was given. We call this a *frequency distribution* because it tells us how frequent each of the possible values is within our sample.

This shows us the absolute frequency of the two responses, for everyone who actually gave a response. We can see from this that there are more people saying "Yes" than "No," but it can be hard to tell from absolute numbers how big the difference is in relative terms. For this reason, we often would rather present the data using *relative frequency*, which is obtained by dividing each frequency by the sum of all frequencies:

$$relative\ frequency_i = \frac{absolute\ frequency_i}{\sum_{j=1}^{N} absolute\ frequency_j}$$

The relative frequency provides a much easier way to see how big the imbalance is. We can also interpret the relative frequencies as percentages by multiplying them by 100. In

Table 3.2. Absolute and Relative Frequencies and Percentages for the *PhysActive* Variable

PhysActive	Absolute frequency	Relative frequency	Percentage
No	2473	0.45	45
Yes	2972	0.55	55

Table 3.3. Frequency Distribution for Number of Hours of Sleep Per Night in the NHANES Dataset

SleepHrsNight	Absolute frequency	Relative frequency	Percentage
2	9	0.00	0.18
3	49	0.01	0.97
4	200	0.04	3.97
5	406	0.08	8.06
6	1172	0.23	23.28
7	1394	0.28	27.69
8	1405	0.28	27.90
9	271	0.05	5.38
10	97	0.02	1.93
11	15	0.00	0.30
12	17	0.00	0.34

this example, we will drop the NA values as well, since we would like to be able to interpret the relative frequencies of active versus inactive people. However, as discussed above we have to assume that the NA values are missing completely at random in order for this to make sense. In this case, if one group of participants was more likely to refuse to answer the question than the other group, then that would *bias* our estimate of the frequency of physical activity, meaning that our estimate would be different from the true value.

Table 3.2 shows us that 45% of the individuals in the NHANES sample said "No" and 55% said "Yes."

Cumulative Distributions

The *PhysActive* variable that we examined above only had two possible values, but often we wish to summarize data that can have many more possible values. When those values are quantitative, one useful way to summarize them is via what we call a *cumulative* frequency representation: rather than asking how many observations take on a specific value, we ask how many observations have that specific value *or less* than that value.

Let's look at another variable in the NHANES dataset, called *SleepHrsNight*, which records how many hours the participant reports sleeping on usual weekdays. Table 3.3 shows a frequency table created as we did above, after removing anyone with missing data

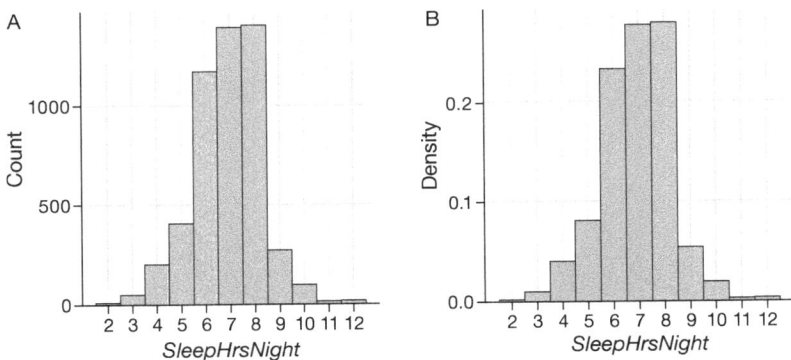

FIGURE 3.2. Histogram showing (A) the number and (B) the relative frequency (also referred to as density or proportion) of people reporting each possible value of the *SleepHrsNight* variable.

for this question. We can already begin to summarize the dataset just by looking at the table; for example, we can see that most people report sleeping between 6 and 8 hours. To see this even more clearly, we can plot a *histogram*, which shows the number of cases having each of the different values (panel A of figure 3.2). We can also plot the relative frequencies, which we often refer to as *densities* (panel B of figure 3.2).

What if we want to know how many people report sleeping five hours or less? To find this, we can compute a *cumulative distribution*. To compute the cumulative frequency for some value *j*, we add up the frequencies for all the values up to and including *j*:

$$cumulative\,frequency_j = \sum_{i=1}^{j} absolute\,frequency_i$$

Let's do this for our sleep variable, computing the absolute and cumulative frequency. In panel A of figure 3.3, we plot the data to see what these representations look like; the absolute frequency values are plotted in solid lines, and the cumulative frequencies are plotted in dashed lines. We see that the cumulative frequency is *monotonically increasing*—that is, it can only go up or stay constant, but it can never decrease. Again, we usually find the relative frequencies to be more useful than the absolute; those are plotted in panel B of figure 3.3. Importantly, the shape of the relative frequency plot is exactly the same as the absolute frequency plot—only the size of the values has changed.

Plotting Histograms

The variables that we examined above were fairly simple, having only a few possible values. Now let's look at a more complex variable: age. Let's start by plotting the *Age* variable for all of the individuals in the NHANES dataset (panel A of figure 3.4). What do you see there? First, you should notice that the number of individuals in each age group is declining over time. This makes sense because the population is being randomly sampled, and thus death

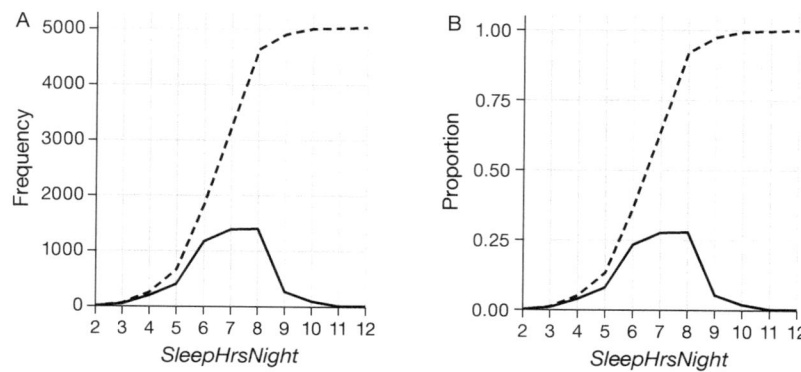

FIGURE 3.3. A plot of the histogram (solid line) and cumulative distribution (dashed line) of absolute frequency (A) and relative frequency (B) for the possible values of *SleepHrsNight*.

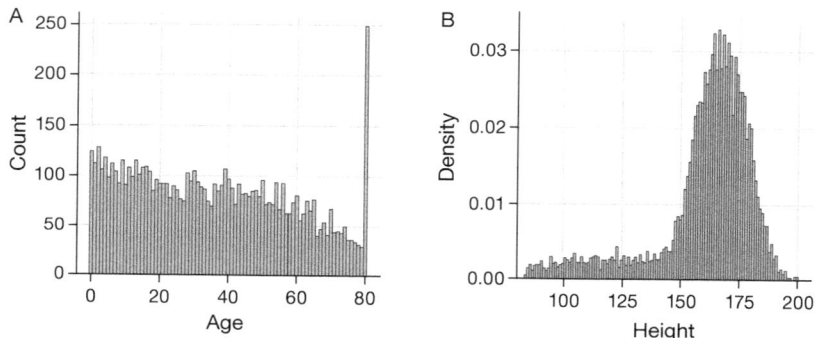

FIGURE 3.4. A histogram of the *Age* (A) and *Height* (B) variables in NHANES.

over time leads to fewer people in the older age ranges. Second, you probably notice a large spike in the graph at age 80. What do you think that's about?

If we were to look up the information about the NHANES dataset, we would see the following definition for the *Age* variable: "Age in years at screening of study participant. Note: Subjects 80 years or older were recorded as 80." The reason for this is that the relatively small number of individuals with very high ages would make it potentially easier to identify the specific person in the dataset if you knew their exact age; researchers generally promise their participants to keep their identity confidential, and this is one of the things they can do to help protect their research subjects. This also highlights the fact that it's always important to know where one's data have come from and how they have been processed; otherwise we might interpret them improperly, thinking that 80-year-olds had been somehow overrepresented in the sample.

Let's look at another more complex variable in the NHANES dataset: height. The histogram of height values is plotted in panel B of figure 3.4. The first thing you should notice about this distribution is that most of its density is centered at about 170 cm, but the

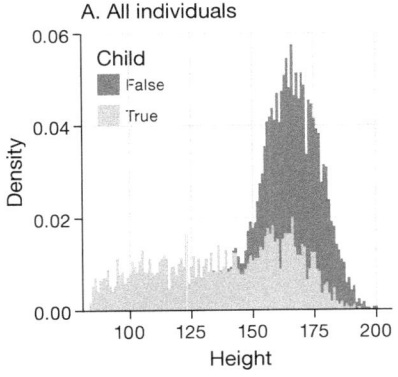
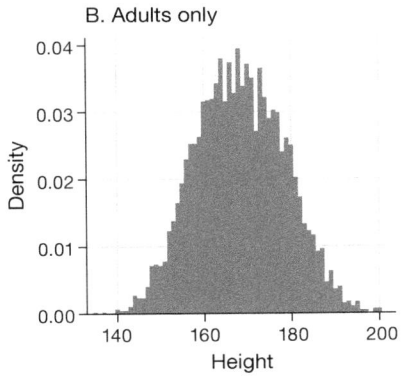

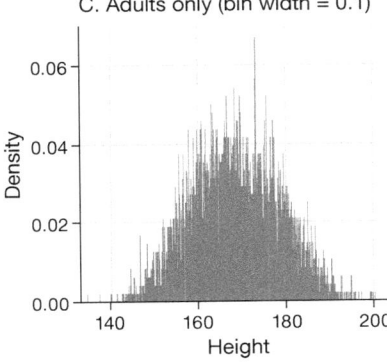

FIGURE 3.5. Histogram of heights for NHANES. (A) Values plotted separately for children (gray) and adults (black). (B) Values for adults only. (C) Same as (B), but with bin width = 0.1.

distribution has a "tail" on the left; there are a small number of individuals with much smaller heights. What do you think is going on here?

You may have intuited that the small heights are coming from the children in the dataset. One way to examine this is to plot the histogram with separate colors for children and adults (panel A of figure 3.5). This shows that all of the very short heights were indeed coming from children in the sample. Let's create a new version of NHANES that only includes adults and then plot the histogram just for them (panel B of figure 3.5). In that plot, the distribution looks much more symmetric. As we will see later, this is a nice example of a *normal* (or *Gaussian*) distribution.

Histogram Bins

In our earlier example with the sleep variable, the data were reported in whole numbers, and we simply counted the number of people who reported each possible value. However, if you look at a few values of the *Height* variable in NHANES (as shown in table 3.4), you will see that it was measured in centimeters down to the first decimal place.

Table 3.4. A Few Values of
the *Height* Variables from
the NHANES Dataset

Height
169.6
169.8
167.5
155.2
173.8
174.5

Panel C of figure 3.5 shows a histogram that counts the density of each possible value down to the first decimal place. That histogram looks really jagged, which is the result of the variability in specific decimal place values. For example, the value 173.2 occurs 32 times, while the value 173.3 occurs only 15 times. There are probably not big differences in the true prevalence of these two heights; more likely this is just due to random variability in our sample of people.

In general, when we create a histogram of data that are continuous or where there are many possible values, we will *bin* the values so that, instead of counting and plotting the frequency of every specific value, we count and plot the frequency of values falling within specific ranges. That's why the plot looks less jagged in panel B of figure 3.5, where we set the bin width to 1; thus, the values 1.3, 1.5, and 1.6 all count toward the frequency of the same bin, which would span from values equal to 1 up through values less than 2.

Once the bin size has been selected, then the number of bins is determined by the data:

$$number\ of\ bins = \frac{range\ of\ scores}{bin\ width}$$

There is no hard-and-fast rule for how to choose the optimal bin width. Occasionally, it will be obvious (as when there are only a few possible values), but in many cases it requires trial and error. There are methods that try to find an optimal bin size automatically, such as the Freedman-Diaconis method that we use in some later examples.

Idealized Representations of Distributions

Datasets are like snowflakes in that every one is different, but nonetheless there are recurring patterns that one often sees across datasets. This allows us to use idealized representations of the data to further summarize them. Let's take the adult height data plotted in figure 3.5 and plot them alongside a very different variable: pulse rate (heartbeats per minute), also measured in NHANES (figure 3.6).

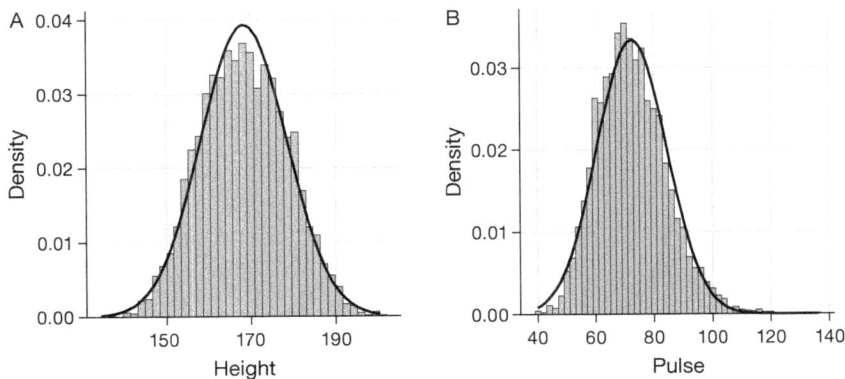

FIGURE 3.6. Histograms for height (A) and pulse (B) in the NHANES dataset, with the normal distribution overlaid for each dataset.

While these plots certainly don't look exactly the same, both have the general characteristic of being relatively symmetric around a rounded peak in the middle. This shape is in fact one of the commonly observed shapes of distributions when we collect data, which we call the *normal* (or *Gaussian*) distribution. This distribution is defined in terms of two values (which we call *parameters* of the distribution): the location of the center peak (which we call the *mean*) and the width of the distribution (which is described in terms of a parameter called the *standard deviation*). Figure 3.6 shows the appropriate normal distribution plotted on top of each of the histograms. You can see that, although the curves don't fit the data exactly, they do a pretty good job of characterizing the distribution—with just two numbers!

As we will see later when we discuss the central limit theorem, there is a deep mathematical reason why many variables in the world exhibit the form of a normal distribution.

Skewness

The examples in figure 3.6 follow the normal distribution fairly well, but in many cases the data will deviate in a systematic way from the normal distribution. One way in which the data can deviate is when they are asymmetric, such that one tail of the distribution is more dense than the other. We refer to this as *skewness*. Skewness commonly occurs when the measurement is constrained to be nonnegative, such as when we are counting things or measuring elapsed times (and thus the variable can't take on negative values).

An example of relatively mild skewness can be seen in the average waiting times at the airport security lines at San Francisco International Airport, plotted in panel A of figure 3.7. You can see that, while most wait times are less than 20 minutes, there are a number of cases where they are much longer—over 60 minutes! This is an example of a "right-skewed" distribution, where the right tail is longer than the left; these are common when looking at counts or measured times, which can't be less than zero. It's less common to see

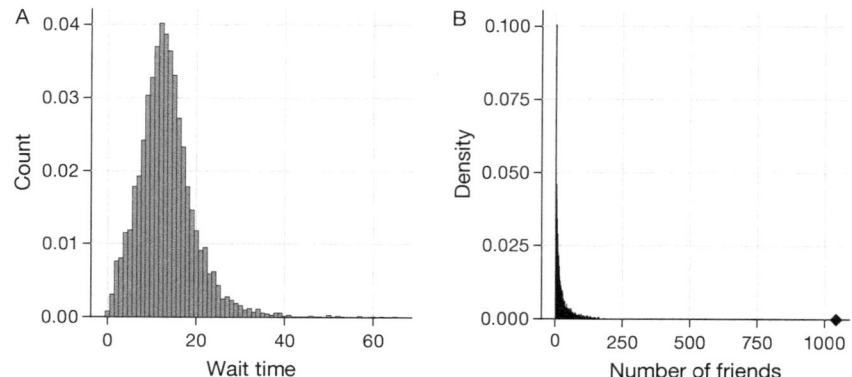

FIGURE 3.7. Examples of right-skewed and long-tailed distributions. (A) Average wait times for security at SFO Terminal A (Jan.–Oct. 2017), obtained from https://awt.cbp.gov/. (B) A histogram of the number of Facebook friends among 3663 individuals, obtained from the Stanford Large Network Database. The person with the maximum number of friends is indicated by the diamond.

"left-skewed" distributions, but they can occur—for example, when looking at fractional values that can't take a value greater than one.

Long-tailed Distributions

Historically, statistics has focused heavily on data that are normally distributed, but there are many data types that look nothing like the normal distribution. In particular, many real-world distributions are "long-tailed," meaning that the right tail extends far beyond the most typical members of the distribution; that is, they are extremely skewed. One of the most interesting types of data where long-tailed distributions occur arises from the analysis of social networks. For an example, let's look at the Facebook friend data from the Stanford Large Network Database and plot the histogram of the number of friends across the 3663 people in the database (panel B of figure 3.7). As we can see, this distribution has a very long right tail—the average person has 24.09 friends, while the person with the most friends (denoted by the diamond) has 1043!

Long-tailed distributions are increasingly being recognized in the real world. In particular, many features of complex systems are characterized by these distributions, from the frequency of words in text to the number of flights in and out of different airports to the connectivity of brain networks. There are a number of different ways that long-tailed distributions can come about, but a common one occurs in cases of the so-called Matthew effect from the Christian Bible:

> For to every one who has will more be given, and he will have abundance; but from him who has not, even what he has will be taken away. (Matthew 25:29, Revised Standard Version)

This is often paraphrased as "the rich get richer." In these situations, advantages compound, such that those with more friends have access to even more new friends, and those with more money have the ability to do things that increase their riches even more.

As the book progresses, we will see several examples of long-tailed distributions, and we should keep in mind that many of the tools in statistics can fail when faced with long-tailed data. As Nassim Nicholas Taleb pointed out in his book *The Black Swan*, such long-tailed distributions played a critical role in the 2008 financial crisis, because many of the financial models used by traders assumed that financial systems would follow the normal distribution, which they clearly did not.

Suggested Reading

- *The Black Swan: The Impact of the Highly Improbable*, by Nassim Nicholas Taleb. This controversial book makes the case that highly significant world events (such as the international financial crisis of 2008) reflect the impact of long-tailed distributions of the sort introduced in this chapter.

Problems

1. Describe why one might think that aggregating data in order to summarize them is a bad idea.
2. Describe the concept of generalization and why it is an important consequence of summarizing data.
3. A dataset contains results from 325 psychology majors, 212 political science majors, 54 economics majors, and 214 students who are undecided. Create a table showing the following values based on these data:
   - Absolute frequency distribution
   - Relative frequency distribution
   - Percentage
4. For the data described in question 3, would it make sense to create a cumulative distribution? Why or why not?
5. A survey of students asks how many siblings each student has, with the following results: 44 with no siblings, 28 with one sibling, 35 with two siblings, 8 with three siblings, and 1 with four siblings. Create a table showing the absolute and cumulative frequencies for these data.
6. Draw a histogram for the absolute frequency data from question 5.
7. What are the two parameters that fully describe a particular normal distribution? Name each parameter and describe what it tells us about the distribution.
8. Describe one common reason that a distribution might be skewed.
9. Describe the concept of a long-tailed distribution and give an example of a variable that you would expect to have such a distribution.

4

Data Visualization

Learning Objectives

Having read this chapter, you should be able to

- Describe and use the principles for distinguishing between good and bad graphs.
- Understand the human limitations that must be accommodated in order to make effective graphs.
- Promise to never create a pie chart. *Ever.*

Why Data Visualization Matters

On January 28, 1986, the space shuttle *Challenger* exploded 73 seconds after takeoff, killing all seven of the astronauts on board. As when any such disaster occurs, there was an official investigation into the cause of the accident, which found that an O-ring connecting two sections of the solid rocket booster leaked, resulting in failure of the joint and explosion of the large liquid fuel tank (figure 4.1).

The investigation found that many aspects of the NASA decision-making process were flawed, and focused in particular on a meeting between NASA staff and engineers from Morton Thiokol, a contractor who built the solid rocket boosters. These engineers were particularly concerned because the temperatures were forecast to be very cold on the morning of the launch, and they had data from previous launches showing that performance of the O-rings was compromised at lower temperatures. In a meeting on the evening before the launch, the engineers presented their data to the NASA managers but were unable to convince them to postpone the launch. Their evidence was a set of handwritten slides showing numbers from various past launches.

The visualization expert Edward Tufte has argued that with a proper presentation of all the data, the engineers could have been much more persuasive. In particular, they could have shown a figure like the one in figure 4.2, which highlights two important facts. First, it shows that the amount of O-ring damage (defined by the amount of erosion and soot found outside the rings after the solid rocket boosters were retrieved from the ocean in previous flights) was closely related to the temperature at takeoff. Second, it shows that

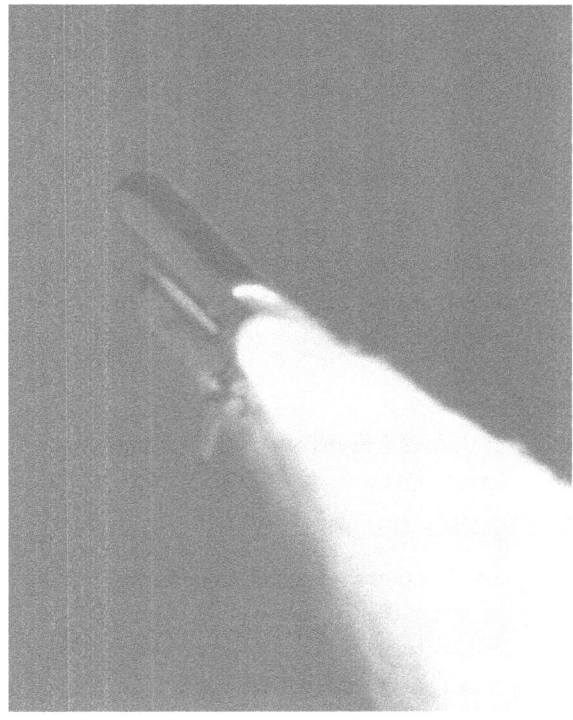

FIGURE 4.1. An image of the solid rocket booster leaking fuel, seconds before the explosion. The small flame visible on the side of the rocket is the site of the O-ring failure. (Photo by NASA; public domain, via Wikimedia Commons)

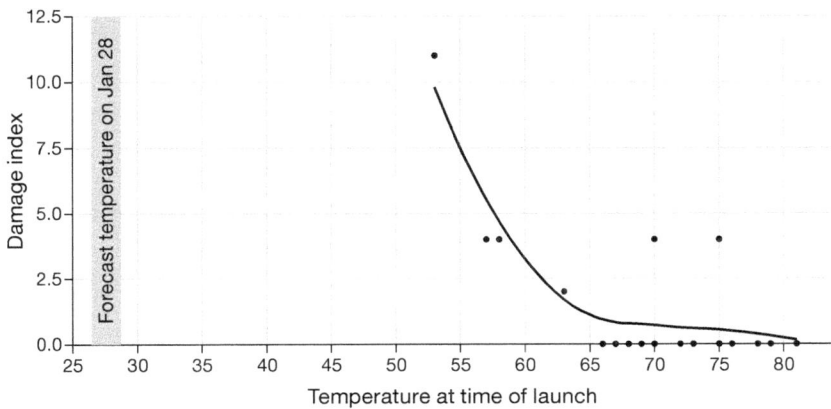

FIGURE 4.2. A replotting of Tufte's damage index data. The line shows the trend in the data, and the shaded patch shows the projected temperatures for the morning of the launch.

the range of forecast temperatures for the morning of January 28 (shown in the shaded area) was well outside the range of all previous launches. While we can't know for sure, it seems at least plausible that this could have been more persuasive.

Anatomy of a Plot

The goal of plotting data is to present a summary of a dataset in a two-dimensional (or occasionally three-dimensional) presentation. We refer to the dimensions as *axes*—the horizontal axis is called the x-axis and the vertical axis is called the y-axis. We can arrange the data along the axes in a way that highlights the data values. These values may be either continuous or categorical.

There are many different types of plots that we can use, each of which has advantages and disadvantages. Let's say that we are interested in characterizing the difference in height between male and female participants in the NHANES dataset.[1] Figure 4.3 shows four different ways to plot these data.

1. The *bar graph* in panel A shows the difference in means, but it doesn't show us how much spread there is in the data around these means—and as we will see later, knowing this is essential to determine whether we think the difference between the groups is large enough to be important. In general, we prefer using a plotting technique that provides a clearer view of the distribution of the data points.
2. The plot in panel B shows the bars with all the data points overlaid using a *beeswarm plot*, which allows us to see the shape of the distribution. This makes it a bit clearer that the distributions of height for men and women are overlapping.
3. In panel C, we see an example of a *violin plot*, which plots the distribution of data in each condition (after smoothing it out a bit).
4. Another option is the *box plot*, shown in panel D, which shows the median (central line), a measure of variability (the width of the box, which is based on a measure called the *interquartile range*), and any outliers (noted by the individual points beyond the ends of the lines).

Items 2, 3, and 4 are each effective ways to show data that provide a good feel for the distribution of the data.

Principles of Good Visualization

Many books have been written on effective visualization of data. There are some principles that most of these authors agree on, while others are more contentious. Here we summarize some of the major principles; if you want to learn more, some good resources are listed in the Suggested Reading section at the end of this chapter.

---

1. The use of *gender* (versus *sex*) as a label for variables throughout this book is based on the labeling in the original dataset.

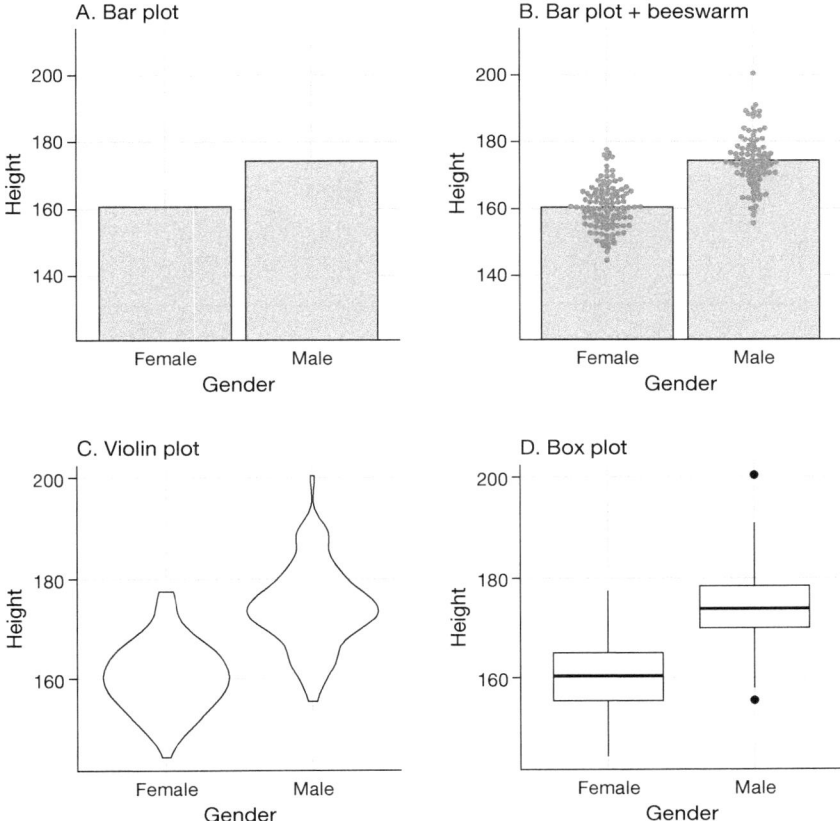

FIGURE 4.3. Four different ways of plotting the difference in height between men and women in a sample from the NHANES dataset. Panel A plots the means of the two groups, which gives no way to assess the relative overlap of the two distributions. Panel B shows the same bars but also overlays the data points using what is known as a beeswarm plot. Panel C shows a violin plot, which shows the distribution of the datasets for each group. Panel D shows a box plot, which highlights the spread of the distribution along with any outliers (shown as individual points).

Show the Data and Make Them Stand Out

Let's say that I performed a study that examined the relationship between dental health and time spent flossing, and I would like to visualize my data. Figure 4.4 shows four possible presentations of these data.

1. In panel A, we don't actually show the data, just a line expressing the relationship between the variables. This is clearly not optimal, because we can't actually see what the underlying data look like. Panels B–D show three possible outcomes from plotting the actual data, where each plot shows a different way that the data might have looked.

2. If we saw the plot in panel B, we would probably be suspicious—rarely would real data follow such a precise pattern.

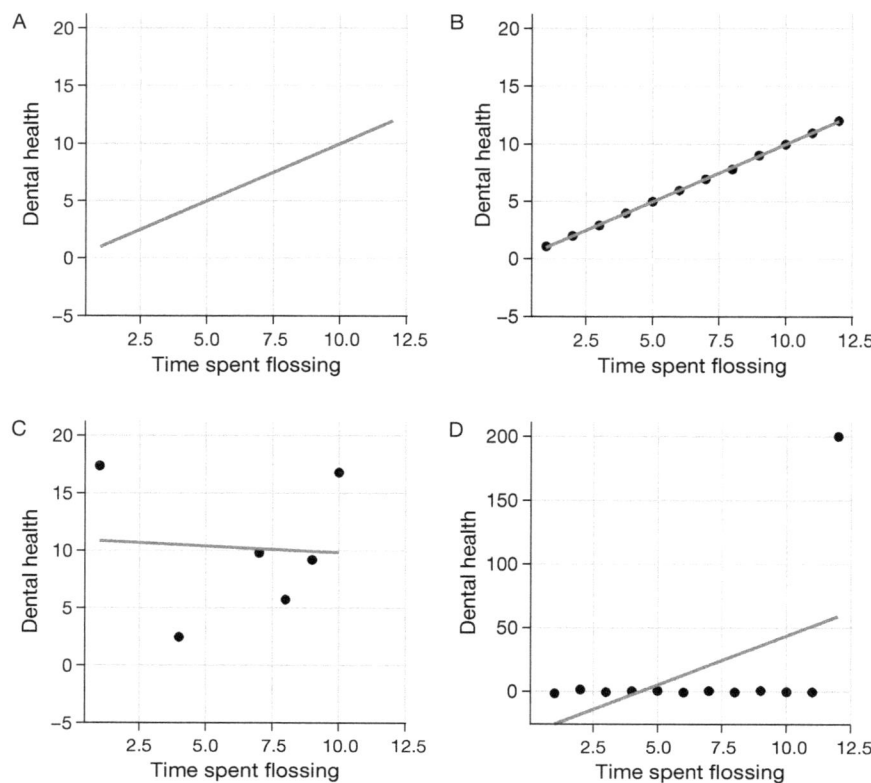

FIGURE 4.4. Four possible presentations of data for the dental health example. Each point in the scatterplot represents one data point in the dataset, and the line in each plot represents the linear trend in the data.

3. The data in panel C, on the other hand, look like real data—they show a general trend, but they are messy, as data in the world usually are.
4. The data in panel D show us that the apparent relationship between the two variables is solely driven by one individual, who we would refer to as an *outlier* because it falls so far outside the pattern of the rest of the group; we discuss outliers in more detail in chapter 17. It should be clear that we probably don't want to conclude very much from an effect that is driven by one data point.

This figure highlights why it is *always* important to look at the raw data before putting too much faith in any summary of the data.

Maximize the Data/Ink Ratio

Edward Tufte has proposed an idea called the *data/ink ratio*:

$$data/ink \; ratio = \frac{amount \; of \; ink \; used \; on \; data}{total \; amount \; of \; ink}$$

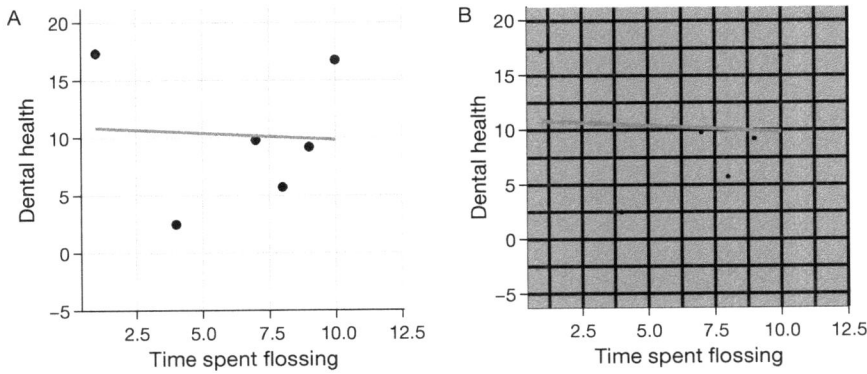

FIGURE 4.5. An example of the same data plotted with two different data/ink ratios.

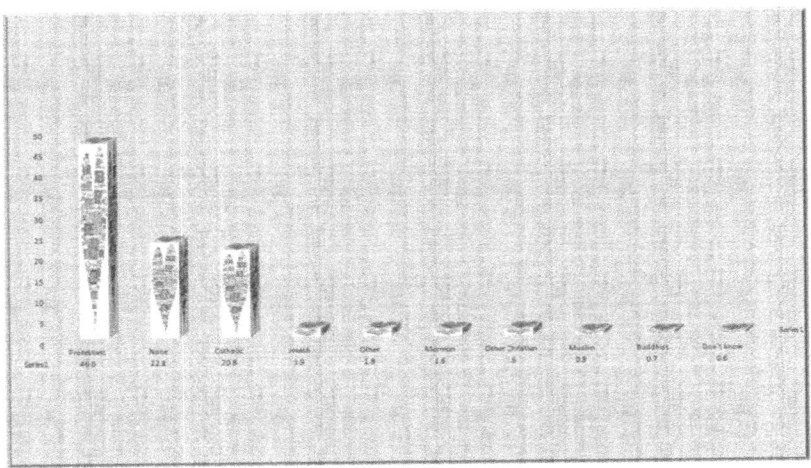

FIGURE 4.6. An example of chartjunk.

The point of this is to minimize visual clutter and let the data show through. For example, take the two presentations of the dental health data in figure 4.5. Both panels show the same data, but panel A is much easier to apprehend because of its relatively higher data/ink ratio.

Avoid Chartjunk

It's especially common to see presentations of data in the popular media that are adorned with lots of visual elements that are thematically related to the content but unrelated to the actual data. This is known as *chartjunk* and should be avoided at all costs.

One good way to avoid chartjunk is to avoid using popular spreadsheet programs to plot one's data. For example, the chart in figure 4.6 (created using Microsoft Excel) plots the relative popularity of different religions in the United States. There are at least three

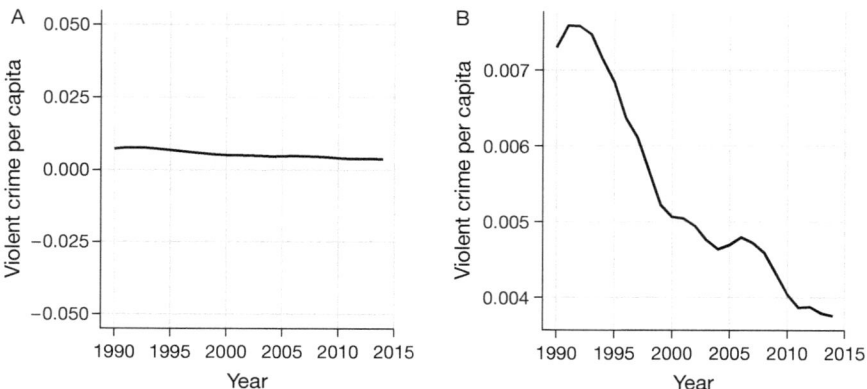

FIGURE 4.7. Crime data from 1990 to 2014 plotted over time. Panels A and B show the same data but with different ranges of values along the y-axis. (Data obtained from https://www.ucrdatatool.gov/Search/Crime/State/RunCrime StatebyState.cfm)

things wrong with this figure:

- It has graphics overlaid on each of the bars that have nothing to do with the actual data.
- It has a distracting background texture.
- It uses three-dimensional bars, which distort the data.

Avoid Distorting the Data

It's often possible to use visualization to distort the message of a dataset. A very common one is use of different axis scaling to either exaggerate or hide a pattern of data. For example, let's say that we are interested in seeing whether rates of violent crime have changed in the US. In figure 4.7, we can see these data plotted in ways that make it look like either crime has remained constant or it has plummeted. The same data can tell two very different stories!

One of the major controversies in statistical data visualization is how to choose the y-axis, and in particular whether it should always include zero. In his famous book *How to Lie with Statistics*, Darrell Huff argued strongly that one should always include the zero point in the y-axis. On the other hand, Edward Tufte has argued against this:

> In general, in a time-series, use a baseline that shows the data not the zero point; don't spend a lot of empty vertical space trying to reach down to the zero point at the cost of hiding what is going on in the data line itself. (https://qz.com/418083/its-ok-not-to -start-your-y-axis-at-zero/)

There are certainly cases where using the zero point makes no sense at all. Let's say that we are interested in plotting body temperature for an individual over time. In figure 4.8, we

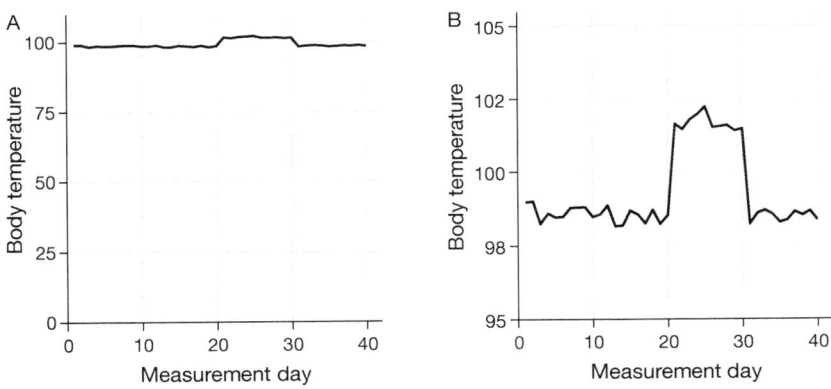

FIGURE 4.8. Body temperature over time, plotted with and without the zero point in the y-axis.

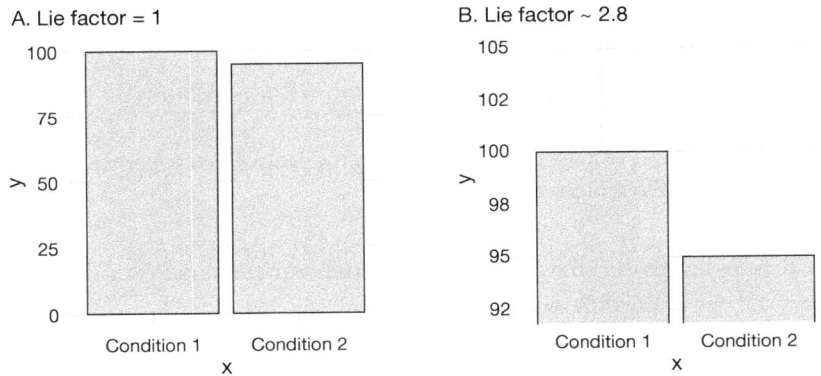

FIGURE 4.9. Two bar charts with associated lie factors.

plot the same (simulated) data with and without zero in the y-axis. It should be obvious that by plotting these data with zero in the y-axis (panel A) we are wasting a lot of space in the figure, given that body temperature of a living person could never go to zero! By including zero, we are also making the apparent jump in temperature during days 21–30 much less evident. In general, my inclination for line plots and scatterplots is to use all the space in the graph, unless the zero point is truly important to highlight.

Edward Tufte introduced the concept of the *lie factor* to describe the degree to which physical differences in a visualization correspond to the magnitude of the differences in the data. If a graphic has a lie factor near one, then it is appropriately representing the data, whereas lie factors far from one reflect a distortion of the underlying data.

The lie factor supports the argument that one should nearly always include the zero point in a bar chart. In figure 4.9, we plot the same data with and without zero in the y-axis. In panel A, the proportional difference in area between the two bars is exactly the same as the proportional difference between the values (i.e., lie factor = 1), whereas in panel B (where zero is not included) the proportional difference in area between the two bars is

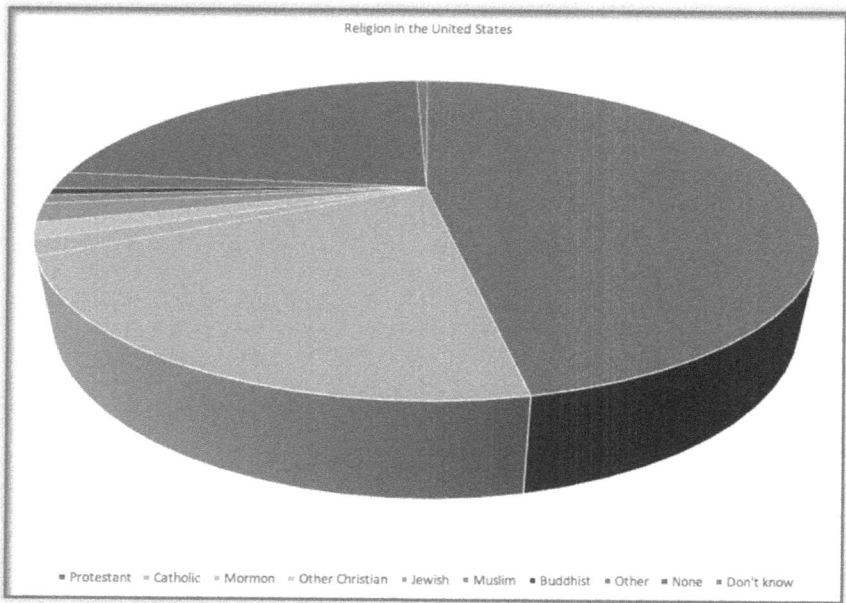

FIGURE 4.10. An example of a pie chart, highlighting the difficulty in apprehending the relative volume of the different pie slices.

roughly 2.8 times bigger than the proportional difference in the values, and thus it visually exaggerates the size of the difference.

Accommodating Human Limitations

Humans have both perceptual and cognitive limitations that can make some visualizations very difficult to understand. It's always important to keep these in mind when building a visualization.

One important perceptual limitation that many people (including myself) suffer from is color blindness. This can make it very difficult to perceive the information in a figure that has only color contrast between the elements but no brightness contrast. It is always helpful to use graph elements that differ substantially in brightness and/or texture, in addition to color. There are also "colorblind-friendly" palettes available for use with many visualization tools.

Even for people with perfect color vision, there are perceptual limitations that can make some plots ineffective. This is one reason why statisticians *never* use pie charts: it can be very difficult for humans to accurately perceive differences in the volume of shapes. The pie chart in figure 4.10 (presenting the same data on religious affiliation that we showed above) shows how tricky this can be.

This plot is terrible for several reasons. First, it requires distinguishing a large number of patterns from very small patches at the bottom of the figure. Second, the visual

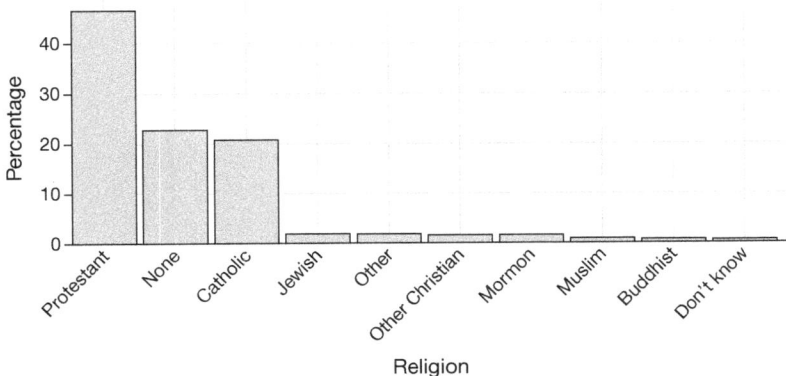

FIGURE 4.11. A clearer presentation of the religious affiliation data.
(Obtained from http://www.pewforum.org/religious-landscape-study/)

perspective distorts the relative numbers, such that the pie wedge for Catholic appears much larger than the pie wedge for None, when in fact the number for None is slightly larger (22.8% versus 20.8%), as is evident in figure 4.6. Third, by separating the legend from the graphic, it requires the viewer to hold information in their working memory in order to map between the graphic and legend and to conduct many "table lookups" in order to continuously match the legend labels to the visualization. And finally, it uses text that is far too small, making it impossible to read without zooming in.

By plotting the data using a more reasonable approach (figure 4.11), we can see the pattern much more clearly. It may not look as flashy as the pie chart generated using Excel, but it's a much more effective and accurate representation of the data. This plot allows the viewer to make comparisons based on the length of the bars along a common scale (the y-axis). Humans tend to be more accurate when decoding differences based on these perceptual elements rather than those based on area or color.

Correcting for Other Factors

Often we want to plot data where the variable of interest is affected by factors other than the one we are interested in. For example, let's say that we want to understand how the price of gasoline has changed over time. Figure 4.12 shows historical gas price data, plotted both with and without adjustment for inflation. Whereas the unadjusted data show a huge increase, the adjusted data show that this is mostly just reflective of inflation. Other examples where one needs to adjust data for other factors include population size and data collected across different seasons.

Suggested Reading

- *Visual Explanations*, by Edward Tufte. This is one of several books by Tufte, who is widely acknowledged as a leading expert on data visualization. His books are beautifully

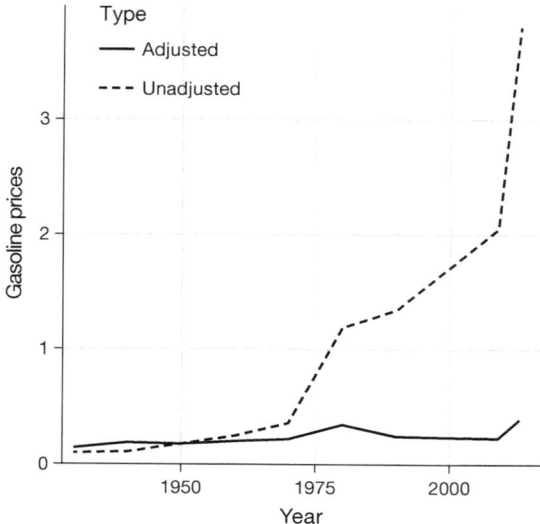

FIGURE 4.12. The price of gasoline in the US
from 1930 to 2013 with and without correction
for inflation (based on the consumer price index).
(Obtained from http://www.thepeoplehistory.com
/70yearsofpricechange.html)

done and provide inspiring examples of how to create data visualizations that are both
visually stunning and informative.

- *Fundamentals of Data Visualization*, by Claus Wilke. Wilke is a computational biologist
  who has written an outstanding, openly available book on how to visualize data using
  the R statistical programming language.

Problems

1. Describe how a better data visualization might have saved the lives of the the space
   shuttle *Challenger* astronauts in 1986.
2. Why might one prefer a plot such as a box plot or violin plot over a simple bar
   graph?
3. Describe the concept of the data/ink ratio. Is a higher ratio better or worse?
4. Outline the arguments for and against always including zero in the y-axis of a plot.
5. Describe the concept of the lie factor and how it relates to the choice of y-axis values
   for a bar graph.
6. Describe at least two psychological reasons that pie charts are problematic.
7. When might one not want to include zero in the y-axis of a plot? Choose all that apply.
   - When doing so obscures important patterns in the data
   - Whenever one plots a bar graph
   - When one wants to make the pattern in the data seem flatter than it actually is

8. Why do we think that normal distributions occur so often in real-world data?
   - They occur when many different factors combine to determine the value of a variable.
   - They occur when the rich get richer.
   - They occur when variables can only take on positive values.
   - They occur when the data measure relationships in a network (such as social networks).
9. Describe the concept of chartjunk and why this kind of visualization is problematic.
10. Why are pie charts particularly bad for data visualization?
    - They rely on our ability to perceive the relative volume.
    - They rely on our ability to distinguish colors.
    - They do not allow correction for factors such as inflation or population size.
    - They can require the viewer to remember colors in order to match the figure to the legend.

5

Fitting Models to Data

One of the fundamental activities in statistics is creating models that can summarize data using a small set of numbers, thus providing a compact description of the data along with a measure of our uncertainty about that description. In this chapter, we discuss the concept of a statistical model and how it can be used to describe data.

Learning Objectives

Having read this chapter, you should be able to

- Describe the basic equation for statistical models (data = model + error).
- Describe different measures of central tendency and dispersion, how they are computed, and which are appropriate under what circumstances.
- Compute a Z-score and describe why they are useful.

What Is a Model?

In the physical world, *Models* are generally simplifications of things in the real world that nonetheless convey the essence of the thing being modeled. A model of a building conveys the structure of the building while being small and light enough to pick up with one's hands; a model of a cell in biology is much larger than the actual thing, but again conveys the major parts of the cell and their relationships.

In statistics, a model is meant to provide a similarly condensed description, but for data rather than for a physical structure. Like physical models, a statistical model is generally much simpler than the data being described; it is meant to capture the structure of the data as simply as possible. In both cases, we realize that the model is a convenient fiction that necessarily glosses over some of the details of the actual thing being modeled. As the statistician George Box famously said: "All models are wrong but some are useful." It can also be useful to think of a statistical model as a theory of how the observed data were generated; our goal then becomes to find the model that most efficiently and accurately summarizes the way in which the data were actually generated. But as we will see below, the desires of efficiency and accuracy are often diametrically opposed to one another.

The basic structure of a statistical model is

$$data = model + error$$

This expresses the idea that the data can be broken into two portions: one portion that is described by a statistical model, which expresses the values that we expect the data to take given our knowledge; and another portion that we refer to as the *error*, which reflects the difference between the model's predictions and the observed data.

In essence, we would like to use our model to predict the value of the data for any given observation. We would write the equation like this:

$$\widehat{data}_i = model_i$$

The "hat" over the data denotes that it's an estimate rather than the actual value of the data. This means that the predicted value of the data for observation i is equal to the value of the model for that observation. Once we have a prediction from the model, we can then compute the error:

$$error_i = data_i - \widehat{data}_i$$

That is, the error for any observation i is the difference between the observed value of the data and the predicted value of the data from the model.

Statistical Modeling: An Example

Let's look at an example of building a model for data, using the data from NHANES. In particular, we will try to build a model of the height of children in the NHANES sample. First, let's load the data and plot them (figure 5.1).

Remember that we want to describe the data as simply as possible while still capturing their important features. The simplest model that we can imagine would involve only a single number; that is, the model would predict the same value for each observation, regardless of what else we might know about those observations. We generally describe a model in terms of its *parameters*, which are values that we can change in order to modify the predictions of the model. Throughout the book we refer to these using the Greek letter beta (β); when the model has more than one parameter, we use subscripted numbers to denote the different betas (e.g., β_1). It's also customary to refer to the values of the data using the letter y, and to use a subscripted version (y_i) to refer to the individual observations.

We generally don't know the true values of the parameters, so we have to estimate them from the data. For this reason, we generally put a "hat" over the β symbol to denote that we are using an estimate of the parameter value rather than its true value. Thus, our simple model for height using a single parameter would be

$$y_i = \hat{\beta} + \epsilon$$

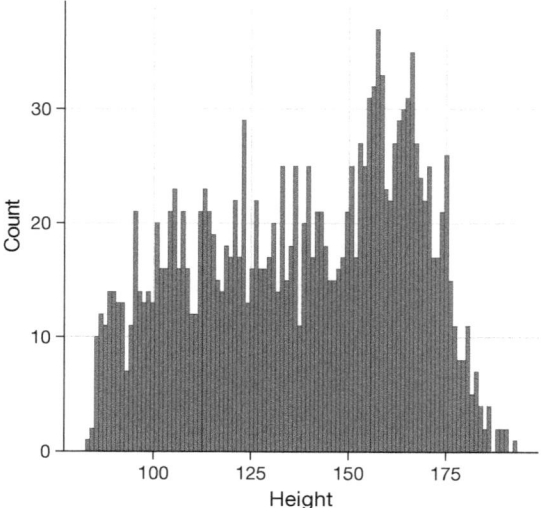

FIGURE 5.1. Histogram of height of children in NHANES.

The subscript i doesn't appear on the right side of the equation, which means that the prediction of the model doesn't depend on which observation we are looking at—it's the same for all of them. The question then becomes, How do we estimate the best values of the parameter(s) in the model? In this particular case, what single value is the best estimate for β? And, more importantly, how do we even define *best*?

One very simple estimator that we might imagine is the *mode*, which is simply the most common value in the dataset. This redescribes the entire set of 1691 children in terms of a single number. If we wanted to predict the height of any new children, then our predicted value would be the same number:

$$\hat{y}_i = 166.5 \ for \ all \ i$$

The error for each individual would then be the difference between the predicted value ($\hat{y}_i$) and their actual height (y_i):

$$error_i = y_i - \hat{y}_i$$

How good of a model is this? In general, we define the goodness of a model in terms of the magnitude of the error, which represents the degree to which the data diverge from the model's predictions; all things being equal, the model that produces lower error is the better model. (Though, as we will see later, all things are usually not equal . . . .) What we find in this case is that the average individual has a fairly large error of -28.8 centimeters when we use the mode as our estimator for β, which doesn't seem very good on its face.

How might we find a better estimator for our model parameter? We might start by trying to find an estimator that gives us an average error of zero. One good candidate is the

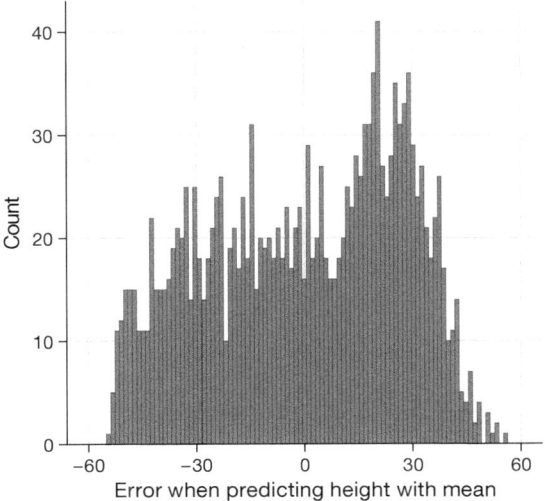

FIGURE 5.2. Distribution of errors from the mean.

arithmetic mean (that is, the *average*, often denoted by a bar over the variable, such as $\bar{X}$), computed as the sum of all the values divided by the number of values. Mathematically, we express this as

$$\bar{X} = \frac{\sum_{i=1}^{n} x_i}{n}$$

It turns out that, if we use the arithmetic mean as our estimator, then the average error is mathematically guaranteed to be zero (see the simple proof at the end of the chapter if you are interested). Even though the average of errors from the mean is zero, we can see from the histogram in figure 5.2 that each individual still has some degree of error; some are positive and some are negative, and those cancel each other out to give an average error of zero.

The fact that the negative and positive errors cancel each other out implies that two different models could have errors of very different magnitude in absolute terms, but they would still have the same average error. This is exactly why the average error is not a good criterion for our estimator; we want a criterion that tries to minimize the overall error regardless of its direction. For this reason, we generally summarize errors in terms of some kind of measure that counts both positive and negative errors as bad. We could use the absolute value of each error value, but it's more common to use the squared errors, for reasons that we discuss later in the book.

There are several common ways to summarize the squared error that you will encounter at various points in this book, so it's important to understand how they relate to one another. First, we could simply add them up; this is referred to as the *sum of squared errors*. The reason we don't usually use this measure is that its magnitude depends on the number of data points, so it can be difficult to interpret unless we are looking at the same number of observations. Second, we could take the mean of the squared error values, which is referred to as the *mean squared error* (MSE). However, because we square the values before averaging, they are not on the same scale as the original data; they are in centimeters squared. For this reason, it's also common to take the square root of the MSE, which we refer to as the *root mean squared error* (RMSE), so that the error is measured in the same units as the original values (in this example, centimeters).

The mean has a pretty substantial amount of error—any individual data point will be about 27 cm from the mean on average—but it's still much better than the mode, which has a root mean squared error of about 39 cm.

Improving Our Model

Can we imagine a better model? Remember that these data are from all children in the NHANES sample, who vary from 2 to 17 years of age. Given this wide age range, we might expect that our model of height should also include age. Let's plot the data for height against age, to see if this relationship really exists.

The points in panel A of figure 5.3 show individuals in the dataset, and there seems to be a strong relationship between height and age, as we would expect. Thus, we might build a model that relates height to age:

$$\hat{y}_i = \hat{\beta} * age_i$$

where $\hat{\beta}$ is our estimate of the parameter that we multiply by age to generate the model prediction.

You may remember from algebra that a line is defined as follows:

$$y = slope * x + intercept$$

If age is the x variable, then that means our prediction of height from age will be a line with a slope of β and an intercept of zero. To see this, let's plot the best-fitting line on top of the data (panel B in figure 5.3).

Something is clearly wrong with this model, as the line doesn't seem to follow the data very well. In fact, the RMSE for this model (39.16 cm) is actually higher than the model that only includes the mean! The problem comes from the fact that our model only includes age, which means that the predicted value of height from the model must take on a value of zero when age is zero. Even though the data do not include any children with an age of zero, the line is mathematically required to have a y value of zero when x is zero, which explains why the line is pulled down below the younger data points. We can fix this by including an intercept in our model, which basically represents the estimated height when age is equal to zero; even though an age of zero is not plausible in this dataset, this is a mathematical trick that allows the model to account for the overall magnitude of the data. The model is

$$\hat{y} = \hat{\beta}_0 + \hat{\beta}_1 * age_i$$

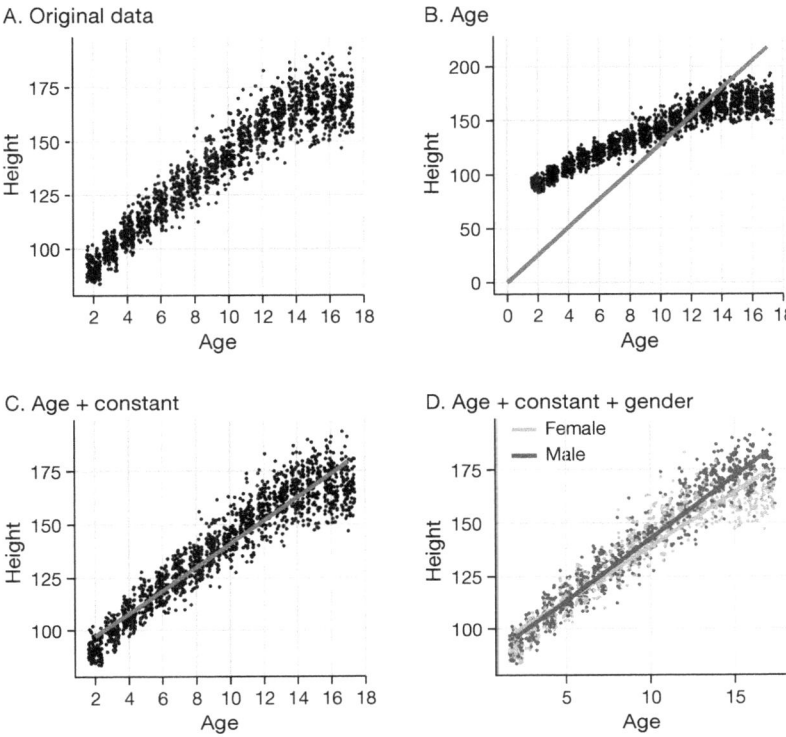

FIGURE 5.3. Height of children in NHANES, plotted without a model (A), with a linear model including only age (B) or age and a constant (C), and with a linear model that fits separate effects of age for males and females (D).

where $\hat{\beta}_0$ is our estimate for the *intercept*, which is a constant value added to the prediction for each individual; we call it the intercept because it maps onto the intercept in the equation for a straight line. We will learn later how it is that we actually estimate these parameter values for a particular dataset; for now, we will use our statistical software to estimate the parameter values that give us the smallest error for these particular data. Panel C in figure 5.3 shows this model applied to the NHANES data, where we see that the line matches the data much better than the one without a constant.

Our error is much smaller using this model—only 8.36 cm on average. Can you think of other variables that might also be related to height? What about gender? In panel D of figure 5.3, we plot the data with lines fitted separately for males and females. From the plot, it seems that there is a difference between boys and girls, but it is relatively small and only emerges after the age of puberty. In figure 5.4, we plot the root mean squared error values across the different models, including one with an additional parameter that models the effect of gender. From this we see that the model got a little bit better going from mode to mean, much better going from mean to mean + age, and only very slightly better by including gender as well.

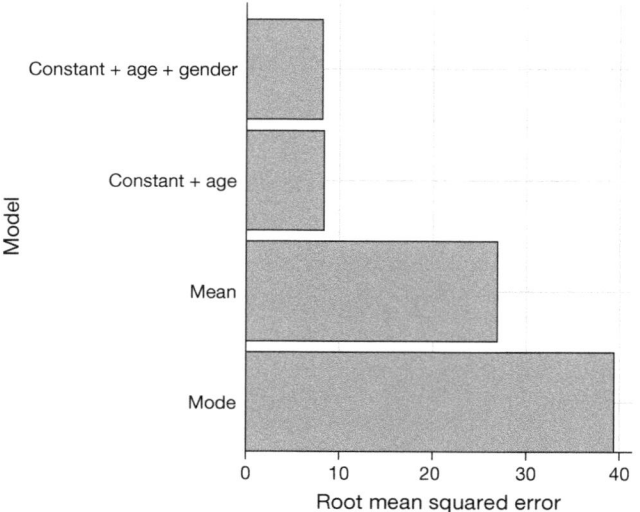

FIGURE 5.4. Mean squared error plotted for each of the models tested above.

What Makes a Model "Good"?

There are generally two things that we want from our statistical model. First, we want it to describe our data well; that is, we want it to have the lowest possible error when modeling our data. Second, we want it to generalize well to new datasets; that is, we want its error to be as low as possible when we apply it to a new dataset in order to make a prediction. It turns out that these two features can often be in conflict.

To understand this, we first need to think about where error comes from. First, it can occur if our model is wrong; for example, if we inaccurately say that height goes down with age instead of going up, then our error will be higher than it would be for the correct model. Similarly, if there is an important factor that is missing from our model, that will also increase our error (as it did when we left age out of the model for height). However, error can also occur even when the model is correct, due to random variation in the data, which we often refer to as *measurement error* or *noise*. Sometimes this really is due to error in our measurement—for example, when the measurements rely on a human, such as using a stopwatch to measure elapsed time in a footrace. In other cases, our measurement device is highly accurate (like a digital scale to measure body weight), but the thing being measured is affected by many different factors that cause it to be variable. If we knew all of these factors, then we could build a more accurate model, but in reality that's rarely possible.

Let's use an example to show this. Rather than using real data, we generate some data using a computer simulation (about which we will have more to say in chapter 8). Let's say that we want to understand the relationship between a person's blood alcohol content (BAC) and their reaction time on a computerized driving test. We can generate some simulated data and plot the relationship (panel A of figure 5.5).

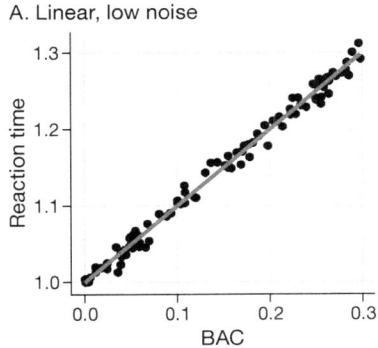

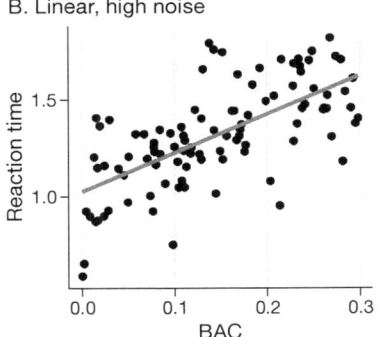

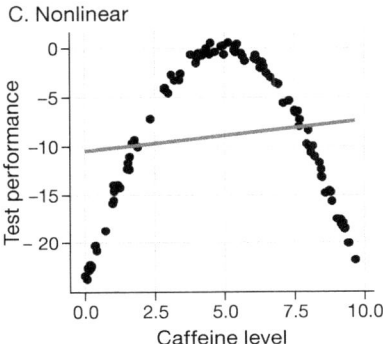

FIGURE 5.5. Simulated relationship between blood alcohol content and reaction time on a driving test, with best-fitting linear model represented by the line. (A) Linear relationship with low measurement error. (B) Linear relationship with higher measurement error. (C) Nonlinear relationship with low measurement error and (incorrect) linear model.

In this example, reaction time goes up systematically with blood alcohol content—the line shows the best-fitting model, and we can see that there is very little error, which is evident from the fact that all the points are very close to the line.

We could also imagine data that show the same linear relationship but have much more error, as in panel B of figure 5.5. Here we see that there is still a systematic increase of reaction time with BAC, but it's much more variable across individuals.

These are both examples where the relationship between the two variables appears to be linear, and the error reflects noise in our measurement. On the other hand, there are situations where the relationship between the variables is not linear, and the error increases because the model is not properly specified. Let's say that we are interested in the relationship between caffeine intake and performance on a test. The relationship between stimulants like caffeine and test performance is often *nonlinear*—that is, it doesn't follow a straight line. This is because performance goes up with smaller amounts of caffeine (as the person becomes more alert) but then starts to decline with larger amounts (as the person becomes nervous and jittery). We can simulate data of this form and then fit a linear model to the data (panel C of figure 5.5). The straight line that best fits these data clearly reflects a high degree of error. Although there is a very lawful relation between test performance and caffeine intake, it follows a curve rather than a straight line. The model, which assumes a linear relationship, has high error because it's the wrong model for these data.

Can a Model Be Too Good?

Error sounds like a bad thing, and usually we prefer a model that has lower error over one that has higher error. However, we mentioned above that there is a tension between the ability of a model to accurately fit the current dataset and its ability to generalize to new datasets, and it turns out that the model with the lowest error often is much worse at generalizing to new datasets!

To see this, let's once again generate some data so that we know the true relation between the variables. We will create two simulated datasets, which are generated in exactly the same way: that is, the equation for both of them is $y = \beta * X + \epsilon$; the only difference is that different random noise was used for ϵ in each case.

We then fit two models to the data: one simple model (with only two parameters, the slope and intercept) and a more complex model that contains a total of eight parameters (the slope and intercept along with size parameters reflecting polynomials of increasing degree, like X^2, X^3, and so on). Panel A in figure 5.6 shows that the more complex (nonlinear) model fits the data better than the simpler (linear) model. However, we see the

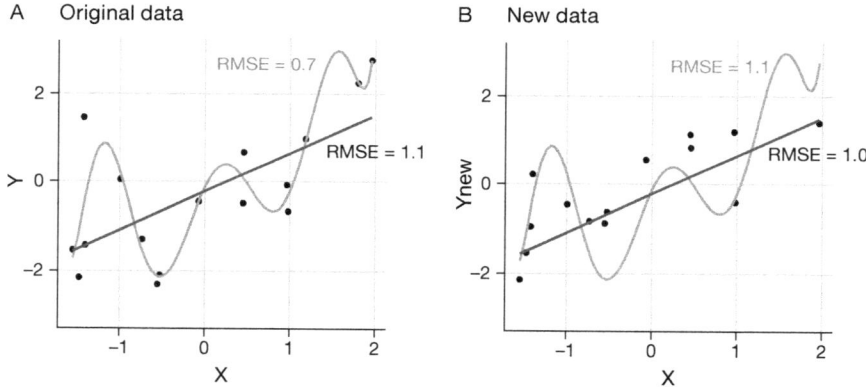

FIGURE 5.6. An example of overfitting. Both datasets were generated using the same model, with different random noise added to generate each set. Panel A shows the data used to fit the model, with a simple linear fit (straight line) and a complex, eighth-order polynomial fit (curved line). The root mean squared error (RMSE) values for each model are shown in the figure; in this case, the complex model has a lower RMSE than the simple model. Panel B shows the second dataset, with the same model overlaid on it and the RMSE values computed using the model obtained from the first dataset. Here we see that the simpler model actually fits the new dataset better than the more complex model, which was overfitted to the first dataset.

opposite when the same model is applied to a new dataset generated in the same way (panel B). Here the simpler model fits the new data better than the more complex model. Intuitively, we can see that the more complex model is influenced heavily by the specific data points in the first dataset; since the exact position of these data points was driven by random noise, this leads the more complex model to fit badly on the new dataset. This is a phenomenon that we call *overfitting*. For now it's important simply to keep in mind that our model fit needs to be good, but not too good. As Albert Einstein (1934) said: "It can scarcely be denied that the supreme goal of all theory is to make the irreducible basic elements as simple and as few as possible without having to surrender the adequate representation of a single datum of experience." Which is often paraphrased as "Everything should be as simple as it can be, but not simpler."

Summarizing Data Using the Mean

We have already encountered the mean (or average) above, and in fact most people know about the average even if they have never taken a statistics class. It is commonly used to describe what we call the *central tendency* of a dataset—that is, what value are the data centered around? Most people don't think of computing a mean as fitting a model to data. However, that's exactly what we are doing when we compute the mean.

We have already seen the formula for computing the mean of a sample of data:

$$\bar{X} = \frac{\sum_{i=1}^{n} x_i}{n}$$

Note that I said this formula was specifically for a *sample* of data, which is a set of data points selected from a larger population. Using a sample, we wish to characterize a larger population—the full set of individuals that we are interested in. For example, if we are political pollsters, our population of interest might be all registered voters, whereas our sample might include just a few thousand people sampled from this population. In chapter 7, we talk in more detail about sampling, but for now the important point is that statisticians generally like to use different symbols to differentiate *statistics* that describe values for a sample from *parameters* that describe the true values for a population; in this case, the formula for the population mean (denoted as μ) is

$$\mu = \frac{\sum_{i=1}^{N} x_i}{N}$$

where N is the size of the entire population. The math is exactly the same for the sample and the population in this case; only the symbols differ. We will later encounter cases where the math is actually different depending on whether we are computing a population parameter or a sample statistic.

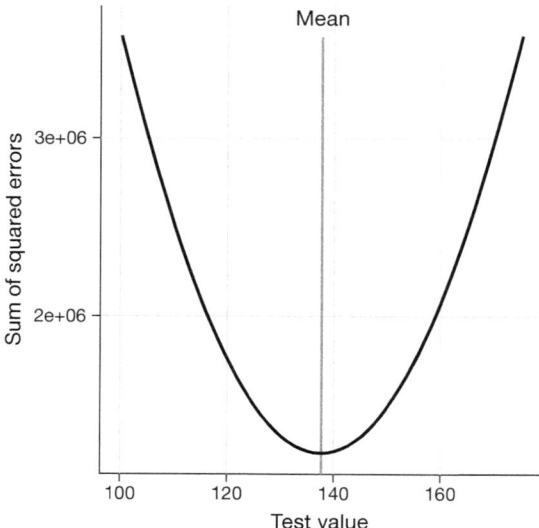

FIGURE 5.7. A demonstration of the mean as the statistic that minimizes the sum of squared errors. Using the NHANES child height data, we compute the mean (denoted by the vertical line). Then we test a range of possible parameter estimates, and for each one we compute the sum of squared errors for each data point from that value, which are denoted by the curve. We see that the mean falls at the minimum of the squared error plot.

We have already seen that the mean is the estimator that is guaranteed to give us an average error of zero, but we also learned that the average error is not the best criterion; instead, we want an estimator that gives us the lowest sum of squared errors (SSE), which the mean also does. One could prove this using calculus, but instead we demonstrate it graphically in figure 5.7.

This minimization of SSE is a good feature, and it's why the mean is the most commonly used statistic to summarize data. However, the mean also has a dark side. Let's say that five people are in a bar and we examine each person's income (table 5.1). The mean ($61,600.00) seems to be a pretty good summary of the income of those five people. Now let's look at what happens if Beyoncé Knowles walks into the bar (table 5.2). The mean is now almost $10 million, which is not really representative of any of the people in the bar—in particular, it is heavily driven by the outlying value of Beyoncé. In general, the mean is highly sensitive to extreme values, which is why it's always important to ensure that there are no extreme values when using the mean to summarize data.

Summarizing Data Robustly Using the Median

If we want to summarize the data in a way that is less sensitive to outliers, we can use another statistic called the *median*. If we were to sort all the values by the order of their magnitude, then the median is the value in the middle. If there is an even number of values,

Table 5.1. Income for Our Five
Bar Patrons

Income	Person
$48,000	Joe
$64,000	Karen
$58,000	Mark
$72,000	Andrea
$66,000	Pat

Table 5.2. Income for Our Five Bar
Patrons Plus Beyoncé Knowles

Income	Person
$48,000	Joe
$64,000	Karen
$58,000	Mark
$72,000	Andrea
$66,000	Pat
$54,000,000	Beyoncé

there will be two values tied for the middle place, in which case we take the mean (i.e., the halfway point) of those two numbers.

Let's look at an example. Say we want to summarize the following values:

8   6   3   14   12   7   6   4   9

If we sort those values:

3   4   6   6   7   8   9   12   14

then the median is the middle value—in this case, the fifth of the nine values.

Whereas the mean minimizes the sum of squared errors, the median minimizes a slighty different quantity: the sum of the *absolute value* of errors. This explains why it is less sensitive to outliers—squaring is going to exacerbate the effect of large errors compared to taking the absolute value. We can see this in the case of the income example: the median income ($65,000) is much more representative of the group as a whole than the mean ($9,051,333) and less sensitive to the one large outlier.

Given this, why would we ever use the mean? As we will see in a later chapter, the mean is the "best" estimator in the sense that it varies less from sample to sample compared to other estimators. It's up to us to decide whether that is worth the sensitivity to potential outliers—statistics is all about trade-offs.

The Mode

Sometimes we wish to describe the central tendency of a dataset that is not numeric. For example, let's say that we want to know which models of iPhone are most commonly used. To test this, we could ask a large group of iPhone users which model each person owns. If we were to take the average of these values, we might see that the mean iPhone model is 9.51, which is clearly nonsensical, since the iPhone model numbers are not meant to be quantitative measurements. In this case, a more appropriate measure of central tendency is the mode, which is the most common value in the dataset, as we discussed above.

Variability: How Well Does the Mean Fit the Data?

Once we have described the central tendency of the data, we often also want to describe how variable the data are. This is sometimes also referred to as *dispersion*, reflecting the fact that it describes how widely dispersed the data are.

We have already encountered the sum of squared errors above, which is the basis for the most commonly used measures of variability: the *variance* and the *standard deviation*. The variance for a population (referred to as σ^2) is simply the sum of squared errors divided by the number of observations—that is, it is exactly the same as the *mean squared error* we encountered earlier:

$$\sigma^2 = \frac{SSE}{N} = \frac{\sum_{i=1}^{n}(x_i - \mu)^2}{N}$$

where μ is the population mean. The population standard deviation is simply the square root of this—that is, the *root mean squared error* that we saw before. The standard deviation is useful because the errors are in the same units as the original data (undoing the squaring that we applied to the errors).

We usually don't have access to the entire population, so we have to compute the variance using a sample, which we refer to as $\hat{\sigma}^2$, with the "hat" representing the fact that this is an estimate based on a sample. The equation for $\hat{\sigma}^2$ (also sometimes called S^2) is similar to the one for σ^2:

$$\hat{\sigma}^2 = \frac{\sum_{i=1}^{n}(x_i - \bar{X})^2}{n - 1}$$

The only difference between the two equations is that we divide by $n - 1$ for the same variance instead of N for the population variance. This relates to a fundamental statistical concept: *degrees of freedom*. Remember that in order to compute the sample variance, we first had to estimate the sample mean $\bar{X}$. Having estimated this, one value in the data is no longer free to vary. For example, let's say we have the following data points for a variable x: $[3, 5, 7, 9, 11]$, the mean of which is 7. Because we know that the mean of this dataset is 7, we can compute what any specific value would be if it were missing. For example, let's

Table 5.3. Variance Estimates using n versus $n - 1$

Estimate	Value
Population variance	725
Variance estimate using n	710
Variance estimate using $n - 1$	725

Note: The estimate using $n - 1$ is closer to the population value.

say we were to obscure the first value (3). Having done this, we still know that its value must be 3, because the mean of 7 implies that the sum of all the values is $7 * n = 35$ and $35 - (5 + 7 + 9 + 11) = 3$.

So when we say that we have "lost" a degree of freedom, it means that there is a value that is not free to vary after fitting the model. In the context of the sample variance, if we don't account for the lost degree of freedom, then our estimate of the sample variance will be *biased*, causing us to underestimate the uncertainty of our estimate of the mean.

Using Simulations to Understand Statistics

I am a strong believer in the use of computer simulations to understand statistical concepts, and in later chapters we dig more deeply into their use. Here we introduce the idea by asking whether we can confirm the need to subtract one from the sample size in computing the sample variance.

Let's treat the entire sample of children from the NHANES dataset as our "population." We want to see how well the calculations of sample variance using either n or $n - 1$ in the denominator estimate the variance of this population, across a large number of simulated random samples from the data. We return to the details of how to do this in chapter 8.

The results in table 5.3 show us that the theory outlined above was correct: the variance estimate using $n - 1$ as the denominator is very close to the variance computed on the full data (i.e., the population), whereas the variance computed using n as the denominator is biased (smaller) compared to the true value. As we will see later, underestimating the variance is particularly problematic because it causes us to be overconfident in our statistical decisions.

Z-Scores

Having characterized a distribution in terms of its central tendency and variability, it is often useful to express the individual scores in terms of where they sit with respect to the overall distribution. Let's say that we are interested in characterizing the relative level of crimes across different states, in order to determine whether California is a particularly dangerous place. We can ask this question using data for 2014 from the FBI's Uniform Crime Reporting site. Panel A of figure 5.8 shows a histogram of the number of violent

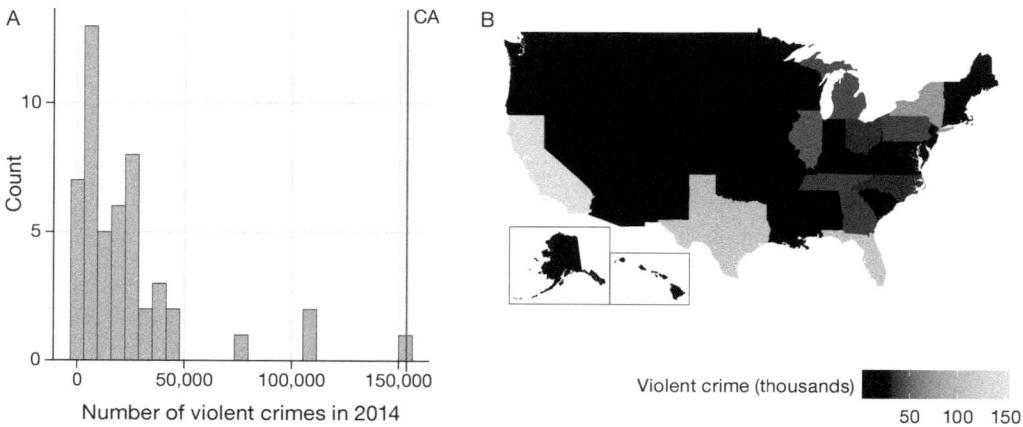

FIGURE 5.8. (A) Histogram of the number of violent crimes. The value for California appears as a vertical line on the right side of the plot. (B) A map of the same data, with number of crimes (in thousands) plotted by state.

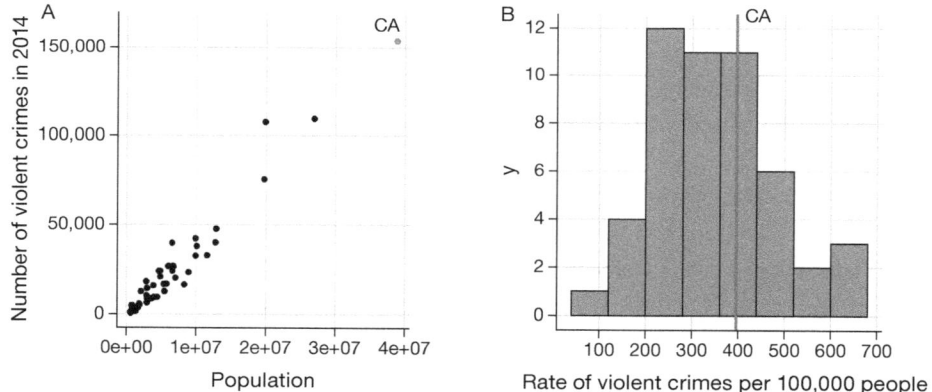

FIGURE 5.9. (A) A plot of the number of violent crimes versus population by state. (B) A histogram of per capita violent crime rates, expressed as crimes per 100,000 people.

crimes per state, highlighting the value for California. Looking at these data, it seems like California is terribly dangerous, with 153,709 crimes in that year. We can visualize these data by generating a map showing the distribution of a variable across states, which is presented in panel B of figure 5.8.

It may have occurred to you, however, that California also has the largest population of any state in the US, so it's reasonable that it would also have a larger number of crimes. If we plot the number of crimes against the population of each state (panel A of figure 5.9), we see that there is a direct relationship between the two variables.

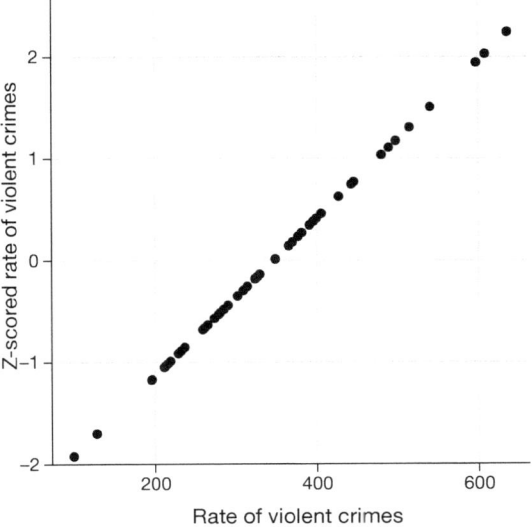

FIGURE 5.10. Scatterplot of original crime rate data against Z-scored data.

Instead of using the raw number of crimes, we should instead use the violent crime *rate* per capita, which we obtain by dividing the number of crimes per state by the population of each state. The dataset from the FBI already includes this value (expressed as rate per 100,000 people). Looking at panel B of figure 5.9, we see that California is not so dangerous after all. Its crime rate of 396.10 per 100,000 people is a bit above the mean of 346.81 across states but well within the range of many other states. But what if we want to get a clearer view of how far it is from the rest of the distribution?

The *Z-score* allows us to express data in a way that provides more insight into each data point's relationship to the overall distribution. The formula to compute a Z-score for an individual data point given that we know the value of the population mean μ and standard deviation σ is

$$Z(x) = \frac{x - \mu}{\sigma}$$

Intuitively, you can think of a Z-score as telling you how far away any data point is from the mean, in units of standard deviation. We can compute this for the crime rate data, as shown in figure 5.10, which plots the Z-scores against the original scores.

The scatterplot shows us that the process of Z-scoring doesn't change the relative distribution of the data points (visible in that the original data and Z-scored data fall on a straight line when plotted against each other)—it just shifts them to have a mean of zero and a standard deviation of one. Figure 5.11 shows the Z-scored crime data using the geographical view. This provides us with a slightly more interpretable look at the data. For example, we can see that Nevada, Tennessee, and New Mexico all have crime rates that are roughly two standard deviations above the mean.

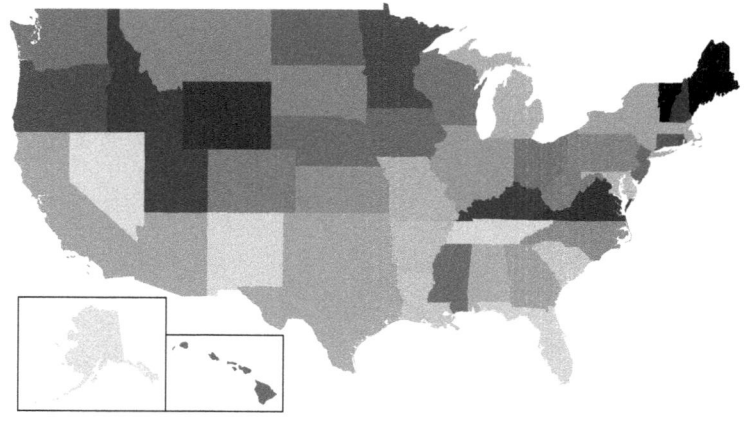

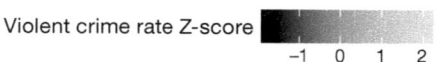

FIGURE 5.11. Crime data rendered onto a US map, presented as Z-scores.

Interpreting Z-scores

The Z in Z-score comes from the fact that the standard normal distribution (that is, a normal distribution with a mean of zero and a standard deviation of one) is often referred to as the Z distribution. We can use the standard normal distribution to help us understand what specific Z-scores tell us about where a data point sits with respect to the rest of the distribution. The left column in figure 5.12 shows that we expect about 16% of values to fall in $Z \geq 1$, and the same proportion to fall in $Z \leq -1$. The right column in figure 5.12 shows the same plot for two standard deviations. Here we see that only about 2.3% of values fall in $Z \leq -2$ and the same in $Z \geq 2$. Thus, if we know the Z-score for a particular data point, we can estimate how likely or unlikely we would be to find a value at least as extreme as that value, which lets us put values into better context. In the case of crime rates, we see that California has a Z-score of 0.38 for its violent crime rate per capita, showing that it is quite near the mean of other states, with about 35% of states having higher rates and 65% of states having lower rates.

Standardized Scores

A *standardized score* is a Z-score that has been transformed to have a mean and standard deviation that is different from the standard normal distribution. Let's say that, instead of Z-scores, we wanted to generate standardized crime scores with a mean of 100 and a standard deviation of 10 (figure 5.13). This is similar to the standardization that is done with scores from intelligence tests to generate the intelligence quotient (IQ). We can do this by simply multiplying the Z-scores by 10 and then adding 100.

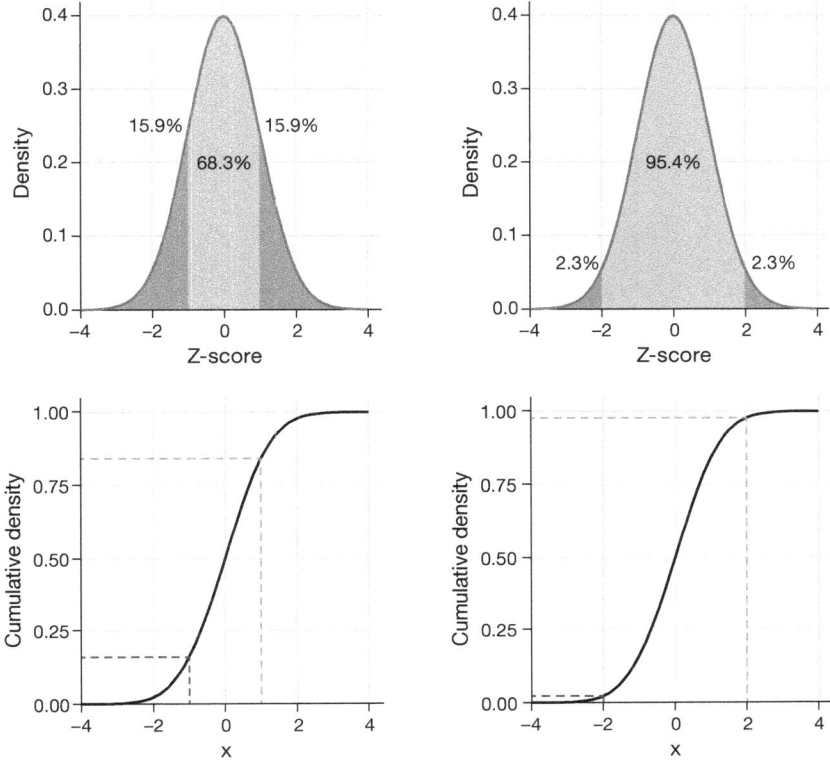

FIGURE 5.12. Density (top) and cumulative distribution (bottom) of a standard normal distribution, with cutoffs at one standard deviation above/below the mean (left column) and two standard deviations (right column).

Using Z-scores to Compare Distributions

One useful application of Z-scores is to compare distributions of different variables. Let's say that we want to compare the distributions of violent crimes and property crimes across states. In panel A of figure 5.14, we plot those against one another, with California shown as an oversize dot. As you can see, the raw rates of property crimes are far higher than the raw rates of violent crimes, so we can't just compare the numbers directly. However, we can plot the Z-scores for these data against one another (panel B of figure 5.14)—here again we see that the distribution of the data does not change. Putting the data into Z-scores for each variable makes them comparable and lets us see that California is actually right in the middle of the distribution in terms of both violent crime and property crime.

Let's add one more factor to the plot: population. In panel A of figure 5.15, we show this using the size of the plotting symbol, which is often a useful way to add information to a plot.

Because Z-scores are directly comparable, we can also compute a *difference score* that expresses the relative rate of violent to nonviolent (property) crimes across states. We can

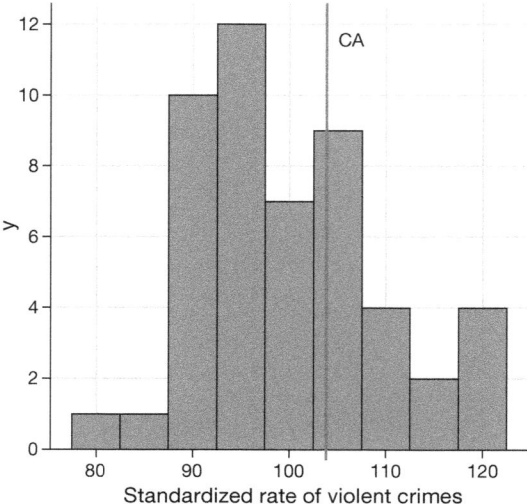

FIGURE 5.13. Crime data presented as standardized scores with mean of 100 and standard deviation of 10.

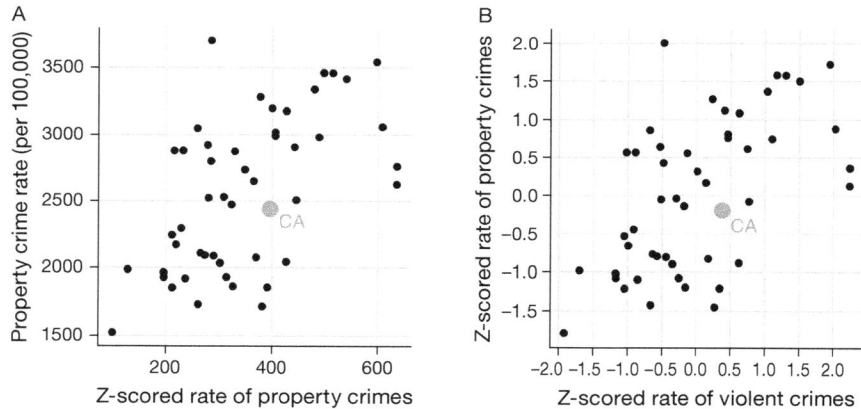

FIGURE 5.14. Plot of violent versus property crime rates (A) and Z-scored rates (B).

then plot those scores against population (panel B of figure 5.15). This shows how we can use Z-scores to bring different variables together on a common scale.

It is worth noting that the smallest states appear to have the largest differences in both directions. While it might be tempting to look at each state and try to determine why it has a high or low difference score, this probably reflects the fact that the estimates obtained from smaller samples are necessarily going to be more variable, as we discuss in chapter 7.

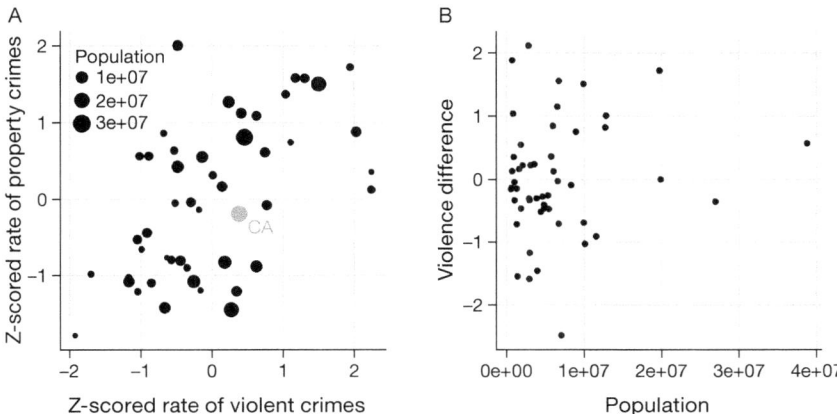

FIGURE 5.15. (A) Plot of violent versus property crime rates, with population size reflected by the size of the plotting symbol. (B) Difference scores for violent versus property crime, plotted against population.

Problems

1. Describe the three parts of the basic model of statistics and how they relate to one another.

2. A researcher wants to create a model to predict the height of individuals, using a sample of eight individuals with the following heights (in centimeters): 170, 176, 168, 188, 178, 168, 179, 181.
   • Determine the mode of these data.
   • Compute the error from the mode for each individual and then compute the average of these errors.
   • Compute the mean of the data and then compute the average error from the mean.

3. Describe the two potential sources of error when comparing a model's predictions to data.

4. Describe the concept of overfitting and how one might know whether overfitting has occurred.

5. If we estimate the mean and the median of a dataset and then compute the sum of squared errors for each of those estimates compared to the data, which of the two estimates is necessarily less than or equal to the other?

6. What is a reason that one might want to use the median rather than the mean to describe a particular dataset?

7. Compute the median of the data described in question 2.

8. What does it mean when a statistical symbol has a "hat" over it (like $\hat{\sigma}$)?

9. What is the difference between the way that the standard deviation is computed for a population versus a sample, and what is the fundamental statistical concept that is related to this difference?

10. Compute the standard deviation for the sample data described in question 2.

11. Compute the Z-scores for each of the individuals described in question 2.
12. Which of the following is true of the mean? Choose all that apply.
    - The sum of errors from each sample to the (sample) mean is zero.
    - It minimizes the sum of squared errors.
    - It is not sensitive to outliers.
    - It reflects the fiftieth percentile in the data.
13. The model that fits a partiuclar dataset best (that is, the one with the lowest sum of squared errors) is generally also the model that fits a new dataset best. True or false?
14. Which of these concepts is most directly relevant to the previous question?
    - Overfitting
    - Degrees of freedom
    - Variability
    - Standardized scores

Appendix

The following is the proof that the sum of errors from the mean is zero:

$$error = \sum_{i=1}^{n}(x_i - \bar{X}) = 0$$

$$\sum_{i=1}^{n} x_i - \sum_{i=1}^{n} \bar{X} = 0$$

$$\sum_{i=1}^{n} x_i = \sum_{i=1}^{n} \bar{X}$$

$$\sum_{i=1}^{n} x_i = n\bar{X}$$

$$\sum_{i=1}^{n} x_i = \sum_{i=1}^{n} x_i$$

6

Probability

Probability theory is the branch of mathematics that deals with chance and uncertainty. It forms an important part of the foundation for statistics, because it provides us with the mathematical tools to describe uncertain events. The study of probability arose in part due to interest in understanding games of chance, like cards or dice. These games provide useful examples of many statistical concepts, because when we repeat these games the likelihood of different outcomes remains (mostly) the same.

There are many deep questions about the meaning of probability that we do not address here; see the Suggested Reading section at the end of this chapter if you are interested in learning more about this fascinating topic and its history.

Learning Objectives

Having read this chapter, you should be able to

- Describe the sample space for a selected random experiment.
- Compute relative frequency and empirical probability for a given set of events.
- Compute probabilities of single events, complementary events, and the unions and intersections of collections of events.
- Describe the law of large numbers.
- Describe the difference between a probability and a conditional probability.
- Describe the concept of statistical independence.
- Use Bayes' theorem to compute the inverse conditional probability.

What Is Probability?

Informally, we usually think of probability as a number that describes the likelihood of some event occurring, which ranges from zero (impossibility) to one (certainty). Sometimes probabilities are instead expressed in percentages, which range from zero to one hundred, as when the weather forecast predicts a 20% chance of rain. In each case, these numbers are expressing how likely that particular event is, ranging from absolutely impossible to absolutely certain.

To formalize probability theory, we first need to define a few terms:

- An *experiment* is any activity that produces or observes an outcome. Examples are flipping a coin, rolling a six-sided die, or trying a new route to work to see if it's faster than the old route.
- The *sample space* is the set of possible outcomes for an experiment. We represent these by listing them within a set of squiggly brackets. For a coin flip, the sample space is {heads, tails}. For a six-sided die, the sample space is each of the possible numbers that can appear: {1, 2, 3, 4, 5, 6}. For the amount of time it takes to get to work, the sample space is all possible real numbers greater than zero (since it can't take a negative amount of time to get somewhere, at least not yet). We won't bother trying to write out all of those numbers within the brackets.
- An *event* is a subset of the sample space. In principle, it could be one or more of the possible outcomes in the sample space, but here we focus primarily on *elementary events*, which consist of exactly one possible outcome. For example, this could be obtaining heads in a single coin flip, rolling a four on a throw of the die, or taking 21 minutes to get home by the new route.

Now that we have those definitions, we can outline the formal features of a probability, which were first defined by the Russian mathematician Andrei Kolmogorov. These are the features that a value *has* to have if it is going to be a probability. Let's say that we have a sample space defined by N independent events, $E_1, E_2, \ldots, E_N$, and X is a random variable denoting which of the events has occurred. $P(X = E_i)$ is the probability of event i:

- Probability cannot be negative: $P(X = E_i) \geq 0$.
- The total probability of all outcomes in the sample space is one; that is, if we take the probability of each E_i and add them up, they must sum to one. We can express this using the summation symbol $\sum$:

$$\sum_{i=1}^{N} P(X = E_i) = P(X = E_1) + P(X = E_2) + \cdots + P(X = E_N) = 1$$

  This is interpreted as saying "Take all of the N elementary events, which we have labeled from one to N, and add up their probabilities. These must sum to one."
- The probability of any individual event cannot be greater than one: $P(X = E_i) \leq 1$. This is implied by the previous point; since they must sum to one and they can't be negative, then any particular probability cannot exceed one.

How Do We Determine Probabilities?

Now that we know what a probability is, how do we actually figure out what the probability is for any particular event?

Personal Belief

Let's say that I asked you what the probability was that Bernie Sanders would have won the 2016 presidential election if he had been the Democratic nominee instead of Hillary Clinton. We can't actually do the experiment to find the outcome. However, most people with knowledge of American politics would be willing to at least offer a guess at the probability of this event. In many cases, personal knowledge and/or opinion is the only guide we have for determining the probability of an event, but this is not very scientifically satisfying.

Empirical Frequency

Another way to determine the probability of an event is to do the experiment many times and count how often each event happens. From the relative frequency of the different outcomes, we can compute the probability of each outcome. For example, let's say that we are interested in knowing the probability of rain in San Francisco. We first have to define the experiment. Let's say that we look at the National Weather Service data for each day in 2017 and determine whether there was any rain at the downtown San Francisco weather station. According to these data, in 2017 there were 73 rainy days. To compute the probability of rain in San Francisco, we simply divide the number of rainy days by the number of days counted (365), giving $P\,(rain\ in\ SF\ in\ 2017) = 0.2$.

How do we know that empirical probability gives us the right number? The answer to this question comes from the *law of large numbers*, which shows that the empirical probability approaches the true probability as the sample size increases. We can see this by simulating a large number of coin flips and looking at our estimate of the probability of heads after each flip. We spend more time discussing simulation in chapter 8; for now, just assume that we have a computational way to generate a random outcome for each coin flip.

Panel A of figure 6.1 shows that, as the number of samples (i.e., coin flip trials) increases, the estimated probability of heads converges onto the true value of 0.5. However, note that the estimates can be very far off from the true value when the sample sizes are small. A real-world example of this was seen in the 2017 special election for the US Senate in Alabama, which pitted the Republican Roy Moore against Democrat Doug Jones. Panel B of figure 6.1 shows the relative amount of the vote reported for each of the candidates over the course of the evening, as an increasing number of ballots were counted. Early in the evening the vote counts were especially volatile, swinging from a large initial lead for Jones to a long period where Moore had the lead, until finally Jones took the lead to win the race.

These two examples show that, while large samples will ultimately converge on the true probability, the results with small samples can be far off. Unfortunately, many people forget this and overinterpret results from small samples. This was referred to as the *law of small numbers* by the psychologists Daniel Kahneman and Amos Tversky, who showed that people (even trained researchers) often behave as if the law of large numbers applies to small samples as well, giving too much credence to results based on small datasets.

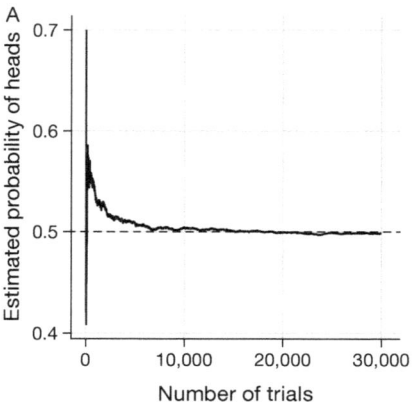
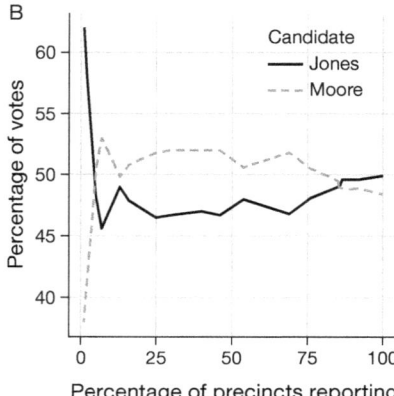

FIGURE 6.1. (A) A demonstration of the law of large numbers. A coin was flipped 30,000 times, and after each flip the probability of heads was computed based on the number of heads and tails collected up to that point. It takes about 15,000 flips for the probability to settle at the true probability of 0.5. (B) Relative proportion of the vote in the December 12, 2017, special election for the US Senate seat in Alabama, as a function of the percentage of precincts reporting. (Data transcribed from https://www.ajc.com/news/national/alabama -senate-race-live-updates-roy-moore-doug-jones/KPRfkdaweoiXICW3FHjXqI/)

We will see examples throughout the book of just how unstable statistical results can be when they are generated on the basis of small samples.

Classical Probability

It's unlikely that any of us has ever flipped a coin tens of thousands of times, but we are nonetheless willing to believe that the probability of flipping heads is 0.5. This reflects the use of yet another approach to computing probabilities, which we refer to as *classical probability*. In this approach, we compute the probability directly based on our knowledge of the mechanism that generates the outcomes.

Classical probability arose from the study of games of chance, such as dice and cards. A famous example comes from a problem encountered by a French gambler who went by the name of Chevalier de Méré. De Méré played two different dice games: in the first he bet on the chance of at least one six on four rolls of a six-sided die, while in the second he bet on the chance of at least one double-six on 24 rolls of two dice. He expected to win money on both of these gambles, but he found that, while on average he won money on the first gamble, he actually lost money on average when he played the second gamble many times. To understand this he turned to his friend, the mathematician Blaise Pascal, who is now recognized as one of the founders of probability theory.

How can we understand this question using probability theory? In classical probability, we start with the assumption that all of the elementary events in the sample space are equally likely; that is, when you roll a die, each of the possible outcomes ($\{1, 2, 3, 4, 5, 6\}$)

is equally likely to occur. (No loaded dice allowed!) Given this, we can compute the probability of any individual outcome as one divided by the number of possible outcomes:

$$P(outcome_i) = \frac{1}{number\ of\ possible\ outcomes}$$

For the six-sided die, the probability of each individual outcome is $1/6$.

This is nice, but de Méré was interested in more complex events, like what happens on multiple dice throws. How do we compute the probability of a complex event (which is a *union* of single events), like rolling a six on the first *or* the second throw? We represent the union of events mathematically using the $\cup$ symbol: for example, if the probability of rolling a six on the first throw is referred to as $P(Roll6_{throw1})$ and the probability of rolling a six on the second throw is $P(Roll6_{throw2})$, then the union is referred to as $P(Roll6_{throw1} \cup Roll6_{throw2})$.

De Méré thought (incorrectly, as we will see below) that he could simply add together the probabilities of the individual events to compute the probability of the combined event, meaning that the probability of rolling a six on the first or second roll would be computed as follows:

$$P(Roll6_{throw1}) = 1/6$$

$$P(Roll6_{throw2}) = 1/6$$

De Méré's incorrect assumption was that

$$P(Roll6_{throw1} \cup Roll6_{throw2}) = P(Roll6_{throw1}) + P(Roll6_{throw2}) = 1/6 + 1/6 = 1/3$$

De Méré reasoned, based on this incorrect assumption, that the probability of at least one six in four rolls was the sum of the probabilities on each of the individual throws: $4 * \frac{1}{6} = \frac{2}{3}$. Similarly, he reasoned that, since the probability of a double-six when throwing two dice is $1/36$, the probability of at least one double-six on 24 rolls of two dice would be $24 * \frac{1}{36} = \frac{2}{3}$. Yet, while he consistently won money on the first bet, he lost money on the second bet. What gives?

To understand de Méré's error, we need to introduce some of the rules of probability theory. The first is the *rule of subtraction*, which says that the probability of some event A not happening is one minus the probability of the event happening:

$$P(\neg A) = 1 - P(A)$$

where $\neg A$ means "not A." This rule derives directly from the axioms that we discussed above; because A and $\neg A$ are the only possible outcomes, then their total probability must sum to 1. For example, if the probability of rolling a one in a single throw is $\frac{1}{6}$, then the probability of rolling anything other than a one is $\frac{5}{6}$.

A second rule tells us how to compute the probability of a conjoint event—that is, the probability that both of two events will occur. We refer to this as an *intersection*, which is signified by the ∩ symbol; thus, $P(A \cap B)$ means the probability that both A and B will occur. We focus now on a version of the rule that tells us how to compute this quantity in the special case when the two events are independent from one another; we will learn later exactly what the concept of *independence* means, but for now we can just take it for granted that the two die throws are independent events. We compute the probability of the intersection of two independent events by simply multiplying the probabilities of the individual events:

$$P(A \cap B) = P(A) * P(B) \text{ if and only if } A \text{ and } B \text{ are independent}$$

Thus, the probability of throwing a six on both of two rolls is $\frac{1}{6} * \frac{1}{6} = \frac{1}{36}$.

The third rule tells us how to add together probabilities—and it is here that we see the source of de Méré's error. The addition rule tells us that, to obtain the probability of either of two events occurring, we add together the individual probabilities but then subtract the likelihood of both occurring together:

$$P(A \cup B) = P(A) + P(B) - P(A \cap B)$$

In a sense, this prevents us from counting those instances twice, and that's what distinguishes the rule from de Méré's incorrect computation. Let's say that we want to find the probability of rolling a six on either of two throws. According to our rules:

$$P(Roll6_{throw1} \cup Roll6_{throw2}) =$$

$$P(Roll6_{throw1}) + P(Roll6_{throw2}) - P(Roll6_{throw1} \cap Roll6_{throw2}) =$$

$$\frac{1}{6} + \frac{1}{6} - \frac{1}{36} = \frac{11}{36}$$

Let's use a graphical depiction to get a different view of this rule. Figure 6.2 shows a matrix representing all possible combinations of results across two throws, and highlights the cells that involve a six on either the first or second throw. If you count up the cells in light gray, you will see that there are 11 such cells. This shows why the addition rule gives a different answer from de Méré's; if we were to simply add together the probabilities for the two throws as he did, then we would count $(6, 6)$ toward both, when it should only be counted once.

Blaise Pascal used the rules of probability to come up with a solution to de Méré's problem. First, he realized that computing the probability of at least one event out of a combination was tricky, whereas computing the probability that something does not occur across several events is relatively easy—it's just the product of the probabilities of the individual events. Thus, rather than computing the probability of at least one six in four

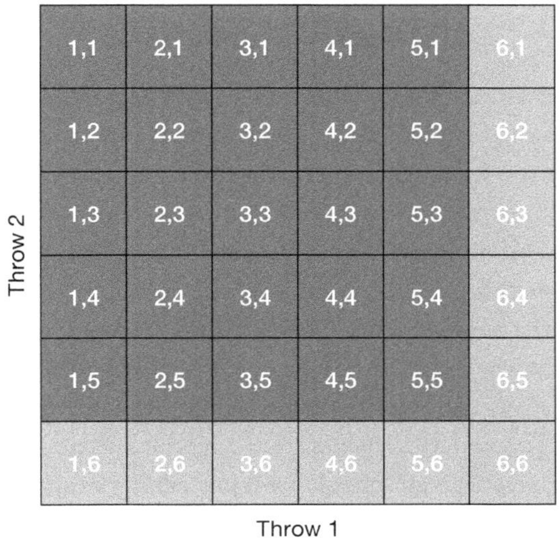

FIGURE 6.2. Each cell in this matrix represents one outcome of two throws of a die, with the columns representing the first throw and the rows representing the second throw. Cells shown in light gray represent the cells with a six in either the first or second throw; the rest are shown in dark gray.

rolls, he instead computed the probability of no sixes across all rolls:

$$P(\textit{no sixes in four rolls}) = \frac{5}{6} * \frac{5}{6} * \frac{5}{6} * \frac{5}{6} = \left(\frac{5}{6}\right)^4 = 0.482$$

He then used the fact that the probability of no sixes in four rolls is the complement of at least one six in four rolls (thus they must sum to one), and used the rule of subtraction to compute the probability of interest:

$$P(\textit{at least one six in four rolls}) = 1 - \left(\frac{5}{6}\right)^4 = 0.517$$

De Méré's gamble that he would throw at least one six in four rolls has a probability of greater than 0.5, explaining why de Méré made money on this bet on average.

But what about de Méré's second bet? Pascal used the same trick:

$$P(\textit{no double-six in 24 rolls}) = \left(\frac{35}{36}\right)^{24} = 0.509$$

$$P(\textit{at least one double-six in 24 rolls}) = 1 - \left(\frac{35}{36}\right)^{24} = 0.491$$

The probability of this outcome was slightly below 0.5, showing why de Méré lost money on average on this bet.

Probability Distributions

A *probability distribution* describes the probability of all the possible outcomes in an experiment. For example, on January 20, 2018, the basketball player Steph Curry hit only two out of four free throws in a game against the Houston Rockets. We know that Curry's overall probability of hitting free throws across the entire season was 0.91, so it seems pretty unlikely that he would hit only 50% of his free throws in a game, but exactly how unlikely is it? We can determine this using a theoretical probability distribution; throughout this book we will encounter a number of probability distributions, each of which is appropriate to describe different types of data. In this case, we use the *binomial* probability distribution, which provides a way to compute the probability of some number of successes out of a number of trials (known as *Bernoulli trials*) in which there is either success or failure and nothing in between, given some known probability of success on each trial. This distribution is defined as

$$P(k; n, p) = P(X = k) = \binom{n}{k} p^k (1 - p)^{n-k}$$

This refers to the probability of k successes on n trials when the probability of success is p. You may not be familiar with $\binom{n}{k}$, which is called the *binomial coefficient*. The binomial coefficient is also referred to as "*n*-choose-*k*" because it describes the number of different ways that one can choose k items out of n total items. The binomial coefficient is computed as

$$\binom{n}{k} = \frac{n!}{k!(n-k)!}$$

where the exclamation point (!) refers to the *factorial* of the number:

$$n! = \prod_{i=1}^{n} i = n * (n-1) * \cdots * 2 * 1$$

The product operator $\prod$ is similar to the summation operator $\sum$, except that it multiplies instead of adds. In this case, it is multiplying together all numbers from one to n.

In the example of Steph Curry's free throws,

$$P(2; 4, 0.91) = \binom{4}{2} 0.91^2 (1 - 0.91)^{4-2} = 0.040$$

This shows that given Curry's overall free throw percentage, it is very unlikely that he would hit only two out of four free throws. Which just goes to show that unlikely things do actually happen in the real world.

Often we want to know not just how likely a specific value is, but how likely it is to find a value that is as extreme as or more than a particular value; this becomes very important

Table 6.1. Simple and Cumulative Probability Distributions for the Number of Successful Free Throws by Steph Curry in Four Attempts

Number of successes	Probability	Cumulative probability
0	0.0001	0.0001
1	0.0027	0.0027
2	0.0402	0.0430
3	0.2713	0.3143
4	0.6857	1.0000

when we discuss hypothesis testing in chapter 9. To answer this question, we can use a *cumulative* probability distribution; whereas a standard probability distribution tells us the probability of some specific value, the cumulative distribution tells us the probability of a value as large or larger (or as small or smaller) than some specific value.

In the free throw example, we might want to know the probability that Steph Curry hits two *or fewer* free throws out of four, given his overall free throw probability of 0.91. To determine this, we could simply use the the binomial probability equation and plug in all the possible values of k and add them together:

$$P(k \leq 2) = P(k=2) + P(k=1) + P(k=0) = 6e^{-5} + .002 + .040 = .043$$

In many cases, the number of possible outcomes is too large for us to compute the cumulative probability by enumerating all possible values; fortunately, it can be computed directly for any theoretical probability distribution. Table 6.1 shows the cumulative probability of each possible number of successful free throws in the example from above, from which we can see that the probability of Curry landing two or fewer free throws out of four attempts is 0.043.

Conditional Probability

So far we have limited ourselves to simple probabilities—that is, the probability of a single event or combination of events. However, we often wish to determine the probability of some event given that some other event has occurred, which are known as *conditional probabilities*.

Let's take the 2016 US presidential election as an example. There are two simple probabilities that we could use to describe the electorate. First, we know the probability that a voter in the US is affiliated with the Republican party: $P(Republican) = 0.44$. We also know the probability that a voter cast their vote in favor of Donald Trump: $P(Trump voter) = 0.46$. However, let's say that we want to know the following: What is the probability that a person cast their vote for Donald Trump *given that they are a Republican*?

To compute the conditional probability of A given B (which we write as $P(A|B)$, "probability of A, given B"), we need to know the *joint probability* (that is, the probability of both

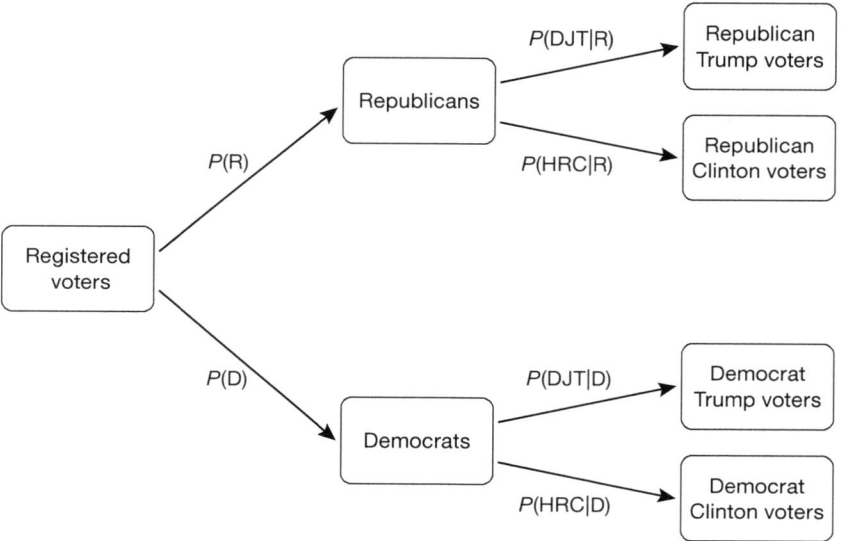

FIGURE 6.3. A graphical depiction of conditional probability, showing how the conditional probability limits our analysis to a subset of the data.

A and *B* occurring) as well as the overall probability of *B*:

$$P(A|B) = \frac{P(A \cap B)}{P(B)}$$

That is, we want to know the probability that both things are true, given that the one being conditioned upon is true.

It can be useful to think of this graphically. Figure 6.3 shows a flowchart depicting how the full population of voters breaks down into Republicans and Democrats, and how the conditional probability (conditioning on party) further breaks down the members of each party according to their vote.

Computing Conditional Probabilities from Data

We can also compute conditional probabilities directly from data. Let's say that we are interested in the following question: What is the probability that someone has diabetes given that they are not physically active—that is, *P(diabetes|inactive)*? The NHANES dataset includes two variables that address the two parts of this question. The first (Diabetes) asks whether the person has ever been told that they have diabetes, and the second (*Phys Active*) records whether the person engages in sports, fitness, or recreational activities that are at least of moderate intensity. Let's first compute the simple probabilities, which are shown in table 6.2. The table shows that the probability that someone in the NHANES dataset has diabetes is 0.1, and the probability that someone is inactive is 0.45.

Table 6.2. Summary Data for Diabetes and Physical Activity

Answer	Number (diabetes)	P(diabetes)	Number (*PhysActive*)	P(*PhysActive*)
No	4893	0.9	2472	0.45
Yes	550	0.1	2971	0.55

Table 6.3. Joint Probabilities for Diabetes and *PhysActive* Variables

Diabetes	*PhysActive*	Number	Probability
No	No	2123	0.39
No	Yes	2770	0.51
Yes	No	349	0.06
Yes	Yes	201	0.04

To compute $P(diabetes|inactive)$, we would also need to know the joint probability of being diabetic *and* inactive, in addition to the simple probabilities of each. These are shown in table 6.3. Based on these joint probabilities, we can compute $P(diabetes|inactive)$. One way to do this in a computer program is to first determine the whether the *PhysActive* variable is equal to "No" for each individual, and then take the mean of those truth values. Since true/false values are treated as 1/0, respectively, by most programming languages (including R and Python), this allows us to easily identify the probability of a simple event by taking the mean of a logical variable representing its truth value. We then use that value to compute the conditional probability, where we find that the probability of someone having diabetes given that they are physically inactive is 0.141.

Independence

The term *independent* has a very specific meaning in statistics, which is somewhat different from the common usage of the word. Statistical independence between two variables means that knowing the value of one variable doesn't tell us anything about the value of the other. This can be expressed as

$$P(A|B) = P(A)$$

That is, the probability of A given some value of B is just the same as the overall probability of A. Looking at it this way, we see that many cases of what we would call "independence" in the real world are not actually statistically independent. For example, there is an ongoing movement by a small group of California citizens to declare a new independent state called Jefferson, which would comprise a number of counties in northern California and Oregon. If this were to happen, the probability that a current California resident

Table 6.4. Proportional Results for Analysis of Mental versus Physical Health

PhysActive	Bad mental health	Good mental health	Total
No	0.10	0.39	0.48
Yes	0.07	0.45	0.52
Total	0.16	0.84	1.00

Table 6.5. Conditional Probabilities for Mental Health given Physical Health

PhysActive	Bad mental health	Good mental health	Total
No	0.20	0.80	1
Yes	0.13	0.87	1
Total	0.16	0.84	1

would now live in the state of Jefferson would be $P(Jeffersonian) = 0.014$, whereas the probability that they would remain a California resident would be $P(Californian) = 0.986$. The new states might be politically independent, but they would *not* be statistically independent, because if we know that a person is Jeffersonian, then we can be sure that they are *not* Californian! That is, while *independence*, in common language, often refers to sets that are exclusive, *statistical independence* refers to instances where one cannot predict anything about one variable from the value of another variable. For example, knowing a person's hair color is unlikely to tell you whether they prefer chocolate or strawberry ice cream.

Let's look at another example, using the NHANES dataset: Are physical health and mental health independent of one another? NHANES includes two relevant questions: *PhysActive*, which asks whether the individual is physically active, and *DaysMentHlth-Bad*, which asks how many days out of the last thirty that the individual experienced bad mental health. Let's consider anyone who recorded more than seven days of bad mental health in the last month to be in bad mental health. Based on this assumption, we can define a new variable called *BadMentalHealth* as a logical variable telling whether each person had more than seven days of bad mental health or not. We can first summarize the data to show how many individuals fall into each combination of the two variables, and then divide by the total number of observations to create a table of proportions (table 6.4).

This shows us the proportion of all observations that fall into each cell. However, what we want to know here is the conditional probability of bad mental health, depending on whether one is physically active or not. To compute this result, we divide each physical activity group by its total number of observations, so that each row now sums to one (shown in table 6.5). Here we see the conditional probabilities of bad or good mental health for each physical activity group (in the top two rows), along with the overall probability of bad or good mental health in the third row. To determine whether mental health and physical activity are independent, we would compare the simple probability of bad mental health (in the third row) to the conditional probability of bad mental health given that one is physically active (in the second row).

The overall probability of bad mental health $P(bad\ mental\ health)$ is 0.16, while the conditional probability $P(bad\ mental\ health|physically\ active)$ is 0.13. Thus, it seems that the conditional probability is somewhat smaller than the overall probability, suggesting that they are not independent—though we can't know for sure just by looking at the numbers, since these numbers might be different due to random variability in our sample. In chapter 12, we discuss statistical tools that let us directly test whether two variables are independent.

Reversing a Conditional Probability: Bayes' Theorem

In many cases, we know $P(A|B)$, but we really want to know $P(B|A)$. This commonly occurs in medical screening, where we know $P(positive\ test\ result|disease)$, but what we want to know is $P(disease|positive\ test\ result)$. For example, at one point it was commonly recommended that men over the age of 50 undergo screening for possible prostate cancer using a prostate-specific antigen (PSA) test. Before a test is approved for use in medical practice, the manufacturer needs to test two aspects of it's performance. First, they need to show how *sensitive* it is—that is, how likely it is to find the disease when it is present:

$$sensitivity = P(positive\ test|disease)$$

Second, they need to show how *specific* it is—that is, how likely it is to give a negative result when there is no disease present:

$$specificity = P(negative\ test|no\ disease)$$

For the PSA test, we know that sensitivity is about 80% and specificity is about 70%. However, these figures don't answer the question that the physician needs to address for any particular patient: What is the likelihood that he actually has cancer, given that the test comes back positive? This requires that we reverse the conditional probability that defines sensitivity: instead of $P(positive\ test|disease)$, we want to know $P(disease|positive\ test)$.

In order to reverse a conditional probability, we can use *Bayes' theorem*:

$$P(B|A) = \frac{P(A|B) * P(B)}{P(A)}$$

Bayes' theorem is fairly easy to derive, based on the rules of probability that we learned earlier in the chapter (see the chapter appendix for this derivation).

If we have only two outcomes, we can express Bayes' rule in a somewhat clearer way, using the sum rule to redefine $P(A)$ by splitting it across the two possible values of B:

$$P(A) = P(A|B) * P(B) + P(A|\neg B) * P(\neg B)$$

Using this, we can redefine Bayes' theorem:

$$P(B|A) = \frac{P(A|B) * P(B)}{P(A|B) * P(B) + P(A|\neg B) * P(\neg B)}$$

We can plug the relevant numbers into this equation to determine the likelihood that an individual with a positive PSA result actually has cancer—but in order to do this, we also need to know the overall probability of cancer for that person, which we often refer to as the *base rate*. Let's take a 60-year-old man, for whom the probability of prostate cancer in the next 10 years is $P(cancer) = 0.058$ based on population data. Using the sensitivity and specificity values that we outlined above, we can compute the individual's likelihood of having cancer given a positive test:

$$P(cancer|test) = \frac{P(test|cancer) * P(cancer)}{P(test|cancer) * P(cancer) + P(test|\neg cancer) * P(\neg cancer)}$$

$$= \frac{0.8 * 0.058}{0.8 * 0.058 + 0.3 * 0.942} = 0.14$$

That's pretty small. Do you find that surprising? Many people do, and in fact there is a substantial psychological literature showing that people systematically neglect *base rates* (i.e., overall prevalence) in their judgments. Confusion about conditional probabilities occurs so often in the courtroom that it has been called the *prosecutor's fallacy*, wherein the probability of the evidence given the defendant's innocence is confused with the probability of innocence given the evidence.

Learning from Data

Another way to think of Bayes' rule is as a way to update our beliefs on the basis of data—that is, learning about the world using data. Let's look at Bayes' rule again:

$$P(B|A) = \frac{P(A|B) * P(B)}{P(A)}$$

The different parts of Bayes' theorem have specific names that relate to their role in using Bayes' theorem to update our beliefs. We start with an initial guess about the probability of B ($P(B)$), which we refer to as the *prior* probability. In the PSA example, we used the base rate in the population as our prior, since it was our best guess as to the individual's chance of cancer before we knew the test result. We then collect some data, which in our example was the test result. The degree to which the data A are consistent with the outcome B is given by $P(A|B)$, which we refer to as the *likelihood*. You can think of this as how likely the data are, given that the particular hypothesis being tested is true. In our example, the hypothesis being tested was whether the individual had cancer, and the likelihood was based on our knowledge about the sensitivity of the test (that is, the probability of a positive test outcome given cancer is present). The denominator $(P(A))$ is referred to as the *marginal likelihood*, because it expresses the overall likelihood of the data, averaged across all the possible values of B (which in our example were disease present and disease absent). The outcome on the left side of the equation $(P(B|A))$ is referred to as the *posterior*, because it's the result at the back end of the computation.

There is another way of writing Bayes' theorem that makes this a bit clearer:

$$P(B|A) = \frac{P(A|B)}{P(A)} * P(B)$$

The value $\frac{P(A|B)}{P(A)}$ tells us how much more or less likely the data A are given B, relative to the overall (marginal) likelihood of the data, while the value $P(B)$ tells us how likely we thought B was before we knew anything about the data. This makes it clearer that the role of Bayes' theorem is to update our prior knowledge based on the degree to which the data are more likely, given B, than they would be overall. If the hypothesis is more likely given the data than it would be in general, then we increase our belief in the hypothesis; if it's less likely given the data, then we decrease our belief.

Odds and Odds Ratios

The result in the last section showed that the likelihood that the individual has cancer based on a positive PSA test is still fairly low, even though it's more than twice as big as it was before we knew the test result. We would often like to quantify the relation between probabilities more directly, which we can do by converting them into *odds*, which express the relative likelihood of something happening or not:

$$odds\ of\ A = \frac{P(A)}{P(\neg A)}$$

In our PSA example, the odds of having cancer (given the positive test) are

$$odds\ of\ cancer = \frac{P(cancer)}{P(\neg cancer)} = \frac{0.14}{1 - 0.14} = 0.16$$

This tells us that the odds of having cancer are fairly low, even though the test is positive. For comparison, the odds of rolling a six in a single dice throw are

$$odds\ of\ 6 = \frac{1}{5} = 0.2$$

As an aside, this is a reason why many medical researchers have become increasingly wary of the use of widespread screening tests for relatively uncommon conditions; most positive results turn out to be false positives, resulting in unneccessary follow-up tests with possible complications, not to mention added stress for the patient.

We can also use odds to compare different probabilities, by computing what is called an *odds ratio*—which is exactly what it sounds like. For example, let's say that we want to know how much the positive test increases the individual's odds of having cancer. We can first compute the *prior odds*—that is, the odds before we knew that the person had tested positively. These are computed using the base rate:

$$prior\ odds = \frac{P(cancer)}{P(\neg cancer)} = \frac{0.058}{1 - 0.058} = 0.061$$

We can then compare these with the *posterior odds*, which are computed using the posterior probability:

$$odds\ ratio = \frac{posterior\ odds}{prior\ odds} = \frac{0.16}{0.061} = 2.62$$

This tells us that the odds of having cancer are increased by 2.62 times given the positive test result. An odds ratio is an example of an *effect size* (discussed further in chapter 10), which is a way of quantifying how relatively large any particular statistical effect is.

What Do Probabilities Mean?

It might strike you that it is a bit odd to talk about the probability of a person having cancer depending on a test result; after all, the person either has cancer or they don't. Historically, there have been two different ways that probabilities have been interpreted. The first (known as the *frequentist* interpretation) interprets probabilities in terms of long-run frequencies. For example, in the case of a coin flip, it would reflect the relative frequencies of heads in the long run after a large number of flips. While this interpretation might make sense for events that can be repeated many times, like a coin flip, it makes less sense for events that only happen once, like an individual person's life or a particular presidential election; and as the economist John Maynard Keynes famously said, "In the long run, we are all dead."

The other interpretation of probablities (known as the *Bayesian* interpretation) is as a degree of belief in a particular proposition. If I were to ask you, "How likely is it that the US will return to the moon by 2040?," you can provide an answer to this question based on your knowledge and beliefs, even though there are no relevant frequencies to compute a frequentist probability. One way that we often frame subjective probabilities is in terms of one's willingness to accept a particular gamble. For example, if you think that the probability of the US landing on the moon by 2040 is 0.1 (i.e., odds of 9 to 1), then that means that you should be willing to accept a gamble that would pay off with anything more than 9 to 1 odds if the event occurs.

As we will see, these two definitions of probability are very relevant to the two different ways that statisticians think about testing statistical hypotheses, which we encounter in later chapters.

Suggested Reading

- *The Drunkard's Walk: How Randomness Rules Our Lives*, by Leonard Mlodinow. Mlodinow is an outstanding popularizer of mathematical ideas, and this book is a fun journey through many of the ideas outlined in this chapter.
- *Ten Great Ideas about Chance*, by Persi Diaconis and Brian Skyrms. Diaconis is one of the world's leading probability theorists, and this book is a fascinating look at some of the major ideas behind probability theory.
- *Ending Medical Reversal: Improving Outcomes, Saving Lives* by Vinayak Prasad and Adam Cifu. This book outlines the history of medical practices (like the prostate-specific antigen test described in this chapter) that have later been recognized as flawed, and discusses the role of statistics in these reversals.

Problems

1. Describe the sample space for each of the following experiments:
   - One roll of an eight-sided die.
   - A marriage proposal.
   - A statistics exam.
2. What are the features that a variable must have in order to be a *probability*?
3. Name and describe the law that justifies the use of data to determine the empirical probability of an event.
4. People often behave as if they can trust probabilities obtained from small samples. What is the name for this psychological phenomenon?
5. Compute the probability of each individual outcome for the following games of chance, assuming that the games are all fair:
   - One roll of an eight-sided die
   - Drawing a particular suit (hearts, clubs, diamonds, or spades) from a standard deck of cards with no jokers
6. If you know that the probablity of having a flight delay on a particular day is 0.12, what is the probability of *not* having a flight delay on that same day? Name the rule that allows you to compute this.
7. Let's say that the probability of catching the common cold in a year is 0.05 and the probability of getting a traffic ticket in a year is 0.03.

- Do you think that these two probabilities are independent?
- Let's assume that they are independent. Compute the probability of both catching a cold and getting a traffic ticket in a year.

8. Describe the rule that is used to add together probabilities. Why can't we just add the two simple probabilities together?

9. For the example in question 7, compute the probability of either catching a cold or getting a ticket in a given year, again assuming that those two probabilities are independent.

10. Describe the concept of a conditional probability and provide a real-world example. What values do you need to know in order to compute it?

11. Describe the concept of statistical independence and its definition, and discuss how it differs from the common use of the term *independent*.

12. The probability of owning a bicycle is 0.40, and the joint probability of owning a bicycle and a car is 0.3. Compute the conditional probability of owning a car given that one owns a bicycle.

13. What is the primary use of Bayes' theorem?

14. Match each of the following terms to its probability definition:
- Sensitivity
- Specificity

15. A medical test has a specificity of 0.8 and a sensitivity of 0.9. A particular patient takes the test and receives a positive result. Previous research has shown that the probability that someone like this patient has the disease is 0.1. Given these values, compute the conditional probability that the patient has the disease given their positive test.

16. What is the difference between the frequentist and Bayesian interpretations of probability?

17. Describe in your own words the difference between a simple probability and a conditional probability. Then give a real-world example of each; do not use one of the examples from the textbook.

18. As of January 20, 2021, Kamala Harris became the first African American vice president of the United States. There have been 49 vice presidents, including Harris. Based on these data, what is the probability that a vice president has been African American?

19. If you know that $P(A|B) = P(A)$, what does this tell you? Choose all that apply.
- A is independent of B.
- Knowing the value of B tells you nothing about the value of A.
- The values of A are necessarily different from the values of B.
- The value of B can be predicted from the value of A.

20. Given the following expression of Bayes' theorem, match each part of the equation with one of the following labels: posterior, prior, likelihood, marginal likelihood.

$$P(Y|X) = \frac{P(X|Y)P(Y)}{P(X)}$$

- $P(Y|X)$
- $P(Y)$
- $P(X|Y)$
- $P(X)$

21. A company has developed a new pregnancy test. Match each of the following concepts to the relevant description: sensitivity, specificity, hit, miss, false alarm, true negative.
    - $P(positive\ test|pregnant)$
    - $P(negative\ test|not\ pregnant)$
    - $P(positive\ test\ AND\ pregnant)$
    - $P(negative\ test\ AND\ pregnant)$
    - $P(positive\ test\ AND\ not\ pregnant)$
    - $P(negative\ test\ AND\ not\ pregnant)$

22. Which of the following must be true if a number is to be a probability? Choose all that apply.
    - It must be greater than zero.
    - It must be greater than or equal to zero.
    - It must be less than one.
    - It must be less than or equal to one.

Appendix

Derivation of Bayes' Theorem

The goal of Bayes' theorem is to allow us to compute $P(B|A)$ from $P(A|B)$, $P(A)$, and $P(B)$. First, remember the rule for computing a conditional probability:

$$P(A|B) = \frac{P(A \cap B)}{P(B)}$$

We can rearrange this to get the formula to compute the joint probability using the conditional:

$$P(A \cap B) = P(A|B) * P(B)$$

Now let's write the formula for the conditional probability of B given A:

$$P(B|A) = \frac{P(B \cap A)}{P(A)}$$

Note that $P(A \cap B)$ is exactly the same as $P(B \cap A)$. Given this, we can simply plug in our formula for $P(A \cap B)$ to compute the inverse probability:

$$P(B|A) = \frac{P(A \cap B)}{P(A)} = \frac{P(A|B) * P(B)}{P(A)}$$

7

Sampling

One of the foundational ideas in statistics is that we can make inferences about an entire population based on a relatively small sample of individuals from that population. In this chapter, we introduce the concept of statistical sampling and discuss why it works.

Learning Objectives

Having read this chapter, you should be able to

- Distinguish between a population and a sample, and between population parameters and sample statistics.
- Describe the concepts of sampling error and sampling distribution.
- Compute the standard error of the mean.
- Describe how the central limit theorem determines the nature of the sampling distribution of the mean.

The Importance of Sampling

Anyone living in the United States will be familiar with the concept of sampling from the political polls that have become a central part of our electoral process. In some cases, these polls can be incredibly accurate at predicting the outcomes of elections. The best-known example comes from the 2008 and 2012 US presidential elections, when the pollster Nate Silver correctly predicted electoral outcomes for 49 of the 50 states in 2008 and for all 50 states in 2012. Silver did this by combining data from 21 different polls, which vary in the degree to which they tend to lean toward either the Republican or Democratic side. Each of these polls included data from about 1000 likely voters—meaning that Silver was able to almost perfectly predict the pattern of votes of more than 125 million voters using data from only about 21,000 people, along with other knowledge (such as how those states have voted in the past).

How Do We Sample?

Our goal in sampling is to estimate the value of some feature of an entire population of interest, using just a small subset of the population. We do this primarily to save time and effort: Why go to the trouble of measuring every individual in the population when a small sample is sufficient to accurately estimate the statistic of interest?

In the election example, the population consists of all registered voters in the region being polled, and the sample is the set of 1000 individuals selected by the polling organization. The way in which we select the sample is critical to ensuring that the sample is *representative* of the entire population, which is a main goal of statistical sampling. It's easy to imagine a nonrepresentative sample; if a pollster only called individuals whose names they had received from the local Democratic party, then it would be unlikely that the results of the poll would be representative of the population as a whole. In general, we define a representative poll as being one in which every member of the population has an equal chance of being selected. When this fails, then we have to worry about whether the statistic that we compute on the sample is *biased*—that is, whether its value is systematically different from the population value (which we refer to as a *parameter*). Keep in mind that we generally don't know this population parameter; if we did, we wouldn't need to sample! But we will use examples where we have access to the entire population, in order to explain some of the key ideas.

It's also important to distinguish between two different ways of sampling: with replacement versus without replacement. In sampling *with replacement*, after a member of the population has been sampled, they are put back into the pool so that they can potentially be sampled again. In sampling without replacement, once a member has been sampled, they are not eligible to be sampled again. It's most common to use sampling without replacement, but there are some contexts in which we use sampling with replacement, as when we discuss a technique called *bootstrapping* in chapter 8.

Sampling Error

Regardless of how representative our sample is, it's likely that the statistic that we compute from the sample is going to differ at least slightly from the population parameter. We refer to this as *sampling error*. If we take multiple samples, the value of our statistical estimate will also vary from sample to sample; we refer to this distribution of our statistic across samples as the *sampling distribution*.

Sampling error is directly related to the quality of our measurement of the population. Clearly, we want the estimates obtained from our sample to be as close as possible to the true value of the population parameter. However, even if our statistic is unbiased (that is, we expect it to have the same value as the population parameter), the value for any particular estimate will differ from the population value, and those differences will be greater when the sampling error is greater. Thus, reducing sampling error is an important step toward better measurement.

Table 7.1. Example Means and
Standard Deviations for Several Samples
of the Height Variable from NHANES

Sample mean	Sample standard deviation
166	11.3
168	9.5
167	9.1
167	9.8
167	10.5

We use the NHANES dataset as an example; we are going to assume that the individuals in the NHANES dataset *are* our population of interest, and then we will draw random samples from this population. We have more to say in the next chapter about exactly how the generation of "random" samples works in a computer.

In this example, we know the adult population mean (168.35) and standard deviation (10.16) for height (measured in centimeters) because we are assuming that the NHANES dataset is the population. Table 7.1 shows the statistics computed from a few samples of 50 individuals from the NHANES population; we can see that, while they are all close to the population value, the statistics computed from each sample vary slightly.

Now let's take a large number of samples (each containing 50 individuals selected at random from the NHANES dataset), compute the mean for each sample, and look at the resulting sampling distribution of means. We have to decide how many samples to take in order to do a good job of estimating the sampling distribution—in this case, we will take 5000 samples so that we are very confident in the answer. Note that simulations like this one can sometimes take a few minutes to run and might make your computer huff and puff. The histogram in figure 7.1 shows that the means estimated for each of the samples of 50 individuals vary somewhat, but that overall they are centered around the population mean. The average of the 5000 sample means (168.3800) is very close to the true population mean (168.3497).

Standard Error of the Mean

Later in the book it will become essential to be able to characterize how variable our samples are, in order to make inferences about the sample statistics. For the mean, we do this by using a quantity called the *standard error* of the mean (SEM), which can be thought of as the standard deviation of the sampling distribution of the mean. To compute the standard error of the mean for our sample, we divide the estimated standard deviation by the square root of the sample size:

$$SEM = \frac{\hat{\sigma}}{\sqrt{n}}$$

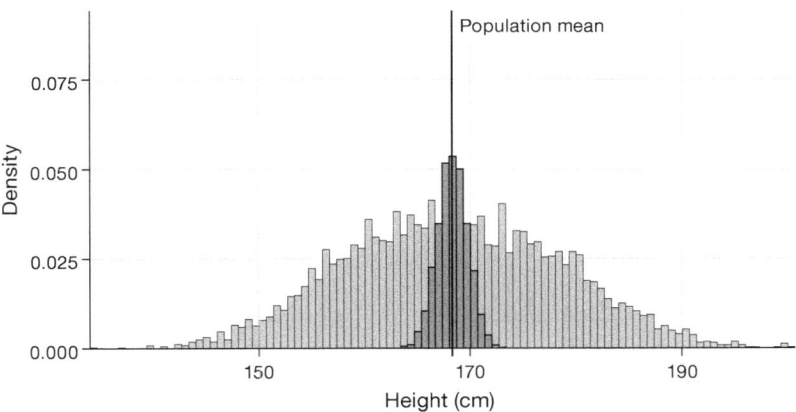

FIGURE 7.1. The dark gray histogram shows the sampling distribution of the mean over 5000 random samples from the NHANES dataset. The histogram for the full dataset is shown in light gray for reference.

Note that we have to be careful about computing SEM using the estimated standard deviation if our sample is small (less than about 30).

Because we have many samples from the NHANES population and we actually know the population SEM (which we compute by dividing the population standard deviation by the size of the population), we can confirm that the SEM computed using the population parameter (1.436) is very close to the observed standard deviation of the means for the samples that we took from the NHANES dataset (1.435).

The formula for the standard error of the mean implies that the quality of our measurement involves two quantities: the population variability and the size of our sample. Because the sample size is the denominator in the formula for SEM, a larger sample size will yield a smaller SEM when holding the population variability constant. We have no control over the population variability, but we *do* have control over the sample size. Thus, if we wish to improve our sample statistics (by reducing their sampling error), then we should use larger samples. However, the formula also tells us something very fundamental about statistical sampling—namely, that the utility of larger samples diminishes with the square root of the sample size. This means that doubling the sample size will *not* double the quality of the statistics; rather, it will improve it by a factor of the square root of two (about 1.41). In chapter 10, we discuss statistical power, which is intimately tied to this idea.

The Central Limit Theorem

The central limit theorem tells us that, as sample sizes get larger, the sampling distribution of the mean will become normally distributed, *even if the data within each sample are not normally distributed.*

First, let's say a little bit about the normal distribution. It's also known as the *Gaussian* distribution, after Carl Friedrich Gauss, a German mathematician who didn't invent it but

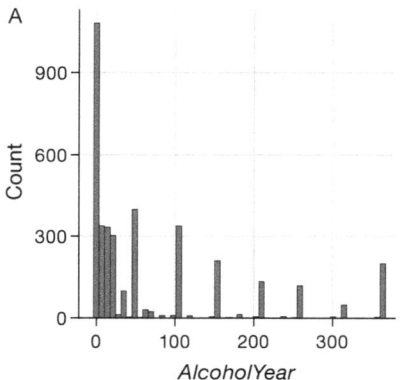
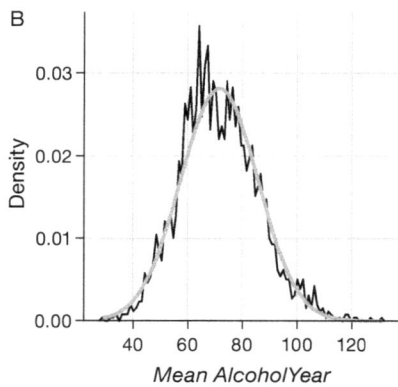

FIGURE 7.2. (A) Distribution of the variable *AlcoholYear* in the NHANES dataset, which reflects the number of days that the individual drank in a year. (B) The sampling distribution of the mean for *AlcoholYear* in the NHANES dataset, obtained by drawing repeated samples of size 50, indicated by the jagged line. The normal distribution with the same mean and standard deviation is shown as a smooth curve.

played a role in its development.[1] The normal distribution is described in terms of two parameters: the mean (which you can think of as the location of the peak) and the standard distribution (which specifies the width of the distribution). The bell-like shape of the distribution never changes, only its location and width. The normal distribution is commonly observed in data collected in the real world, as we have already seen in chapter 3—and the central limit theorem gives us some insight into why that occurs.

To see the central limit theorem in action, let's work with the variable *AlcoholYear* from the NHANES dataset, which is highly skewed, as shown in panel A of figure 7.2. This distribution is, for lack of a better word, funky—and definitely not normally distributed. Now let's look at the sampling distribution of the mean for this variable, shown in panel B of figure 7.2. The sampling distribution is obtained by repeatedly drawing samples of size 50 from the NHANES dataset and taking the mean. Despite the clear nonnormality of the original data, the sampling distribution is remarkably close to the normal distribution.

The central limit theorem is important for statistics because it allows us to safely assume that the sampling distribution of the mean will be normal in most cases. This means that we can take advantage of statistical techniques that assume a normal distribution. It's also important because it tells us why normal distributions are so common in the real world; any time we combine many different factors into a single number, the result is likely to be a normal distribution. For example, the height of any adult depends on a complex mixture of their genetics and experience; even if those individual contributions may not be normally distributed, when we combine them the result is a normal distribution.

1. This is an example of Stigler's law of eponymy, named after the statistician Stephen Stigler, whom we met in chapter 1. This law asserts that scientific discoveries are rarely named after their original discoverer, who in the case of the Gaussian distribution was the French mathematician Abraham de Moivre.

Suggested Reading

- *The Signal and the Noise: Why So Many Predictions Fail—but Some Don't,* by Nate Silver. An outstanding overview of the ideas behind statistical prediction.

Problems

1. What is the definition of a representative sample?
2. Let's say that we want to know the attitudes of citizens in a small town regarding a proposed water conservation plan. A pollster performs polling at three local churches that shows strong support for the proposal.
   - Why might one think that this is not a representative sample?
   - How might the pollster change the polling procedure to make it more representative?
3. Name the concept that describes the fact that statistical estimates vary from sample to sample.
4. The standard error of the mean can be thought of as the _____ of the _____ distribution of the mean.
5. What are the two quantities that determine the quality of a measurement? Which of these do we have control over, and how can we take advantage of that?
6. What does the formula for the standard error of the mean imply about how the quality of a measurement changes as a function of sample size?
7. For this example, we reuse the sample from question 2 in chapter 5 of eight individuals with the following heights (in centimeters): 170, 176, 168, 188, 178, 168, 179, 181. Compute the standard error of the mean.
8. What does the central limit theorem tell us will happen as sample sizes get larger, and why is this so important for statistics?
9. How does the central limit theorem relate to the fact that normal distributions are so often observed in the real world?
10. Which of these is the most appropriate description of the concept of a sampling distribution for a statistic?
    - It is the distribution of that statistic across many samples.
    - It is the distribution of the values obtained within a specific sample.
    - It is the distribution of the entire population from which the samples are obtained.
    - It is the distance of the sample mean from the population mean.
11. To compute the standard error of the mean, one divides _____ by _____.
12. Given these values, compute the standard error of the mean:
    - Mean = 120
    - Variance = 16
    - Sample size = 100

8

Resampling and Simulation

The use of computer simulations has become an essential aspect of modern statistics. For example, one of the most important books in practical computer science, called *Numerical Recipes in C*, says the following:

> Offered the choice between mastery of a five-foot shelf of analytical statistics books and middling ability at performing statistical Monte Carlo simulations, we would surely choose to have the latter skill. (Press et al. 1992, p. 691)

In this chapter, we introduce the concept of a Monte Carlo simulation and discuss how it can be used to perform statistical analyses.

Learning Objectives

Having read this chapter, you should be able to

- Describe the concept of a Monte Carlo simulation.
- Describe the meaning of randomness in statistics.
- Describe how pseudorandom numbers are generated.
- Describe the concept of the bootstrap and its use of resampling.

Monte Carlo Simulation

The concept of Monte Carlo simulation was devised by the mathematicians Stan Ulam and Nicholas Metropolis, who were working to develop an atomic weapon for the US as part of the Manhattan Project. They needed to compute the average distance that a neutron would travel in a substance before it collided with an atomic nucleus, but they could not compute this using standard mathematics. Ulam realized that these computations could be simulated using random numbers, just like a casino game. In a casino game, such as a roulette wheel, numbers are generated at random; to estimate the probability of a specific outcome, one could play the game hundreds of times. Ulam's uncle had gambled at the Monte Carlo casino in Monaco, which is apparently where the name came from for this new technique.

There are four steps to performing a Monte Carlo simulation:

1. Define a domain of possible values.
2. Generate random numbers within that domain from a probability distribution.
3. Perform a computation using the random numbers.
4. Combine the results across many repetitions.

As an example, let's say that I want to figure out how much time to allow for an in-class quiz. We will pretend for the moment that we know that the distribution of quiz completion times is normal, with a mean of five minutes and a standard deviation of one minute. Given this information, how long does the test period need to be so that we can expect all students to finish the exam 99% of the time? There are two ways to solve this problem. The first is to calculate the answer using a mathematical theory known as the statistics of extreme values. However, this involves complicated mathematics. Alternatively, we could use Monte Carlo simulation. To do this, we need to generate random samples from a normal distribution.

Randomness in Statistics

The term *random* is often used colloquially to refer to things that are bizarre or unexpected, but in statistics the term has a very specific meaning: A process is *random* if it is unpredictable. For example, if I flip a fair coin 10 times, the value of the outcome on one flip does not provide me with any information that lets me predict the outcome on the next flip. It's important to note that the fact that something is unpredictable doesn't necessarily mean that it is not deterministic. For example, when we flip a coin, the outcome of the flip is determined by the laws of physics; if we know all the conditions in enough detail, we should be able to predict the outcome of the flip. However, in practice, many factors combine to make the outcome of the coin flip unpredictable.

Psychologists have shown that humans actually have a fairly bad sense of randomness. First, we tend to see patterns when they don't exist. In the extreme, this leads to the phenomenon of *pareidolia*, in which people perceive familiar objects within random patterns (such as perceiving a cloud as a human face or seeing the Virgin Mary in a piece of toast). Second, humans tend to think of random processes as self-correcting, which leads us to expect that we are "due for a win" after losing many rounds in a game of chance, a phenomenon known as the *gambler's fallacy*.

Generating Random Numbers

Running a Monte Carlo simulation requires that we generate random numbers. Generating truly random numbers (i.e., numbers that are completely unpredictable) is only possible through physical processes, such as the decay of atoms or the rolling of dice, which are difficult to obtain and/or too slow to be useful for computer simulation. In general, instead of truly random numbers we use *pseudorandom* numbers generated using a computer

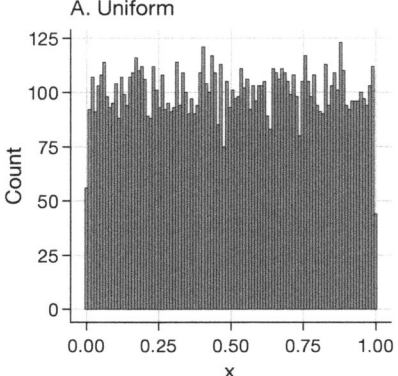
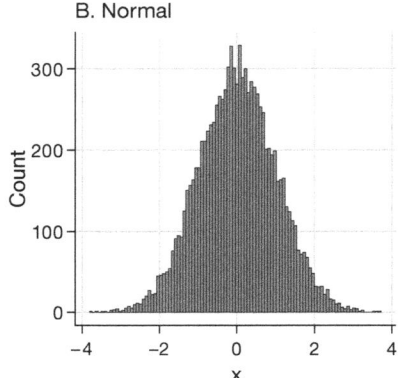

FIGURE 8.1. Examples of histograms of random numbers generated from a uniform (A) or normal (B) distribution.

algorithm; these numbers seem random in the sense that they are difficult to predict, but the series of numbers actually repeats at some point. For example, the random number generator used in R repeats after $2^{19937} - 1$ numbers. That's far more than the number of seconds in the history of the universe, and we generally think that this is fine for most purposes in statistical analysis.

Most statistical software includes functions to generate random numbers for each of the major probability distributions, such as the uniform distribution (all values between 0 and 1 equally distributed), the normal distribution (described in chapter 7), and the binomial distribution (e.g., coin flips). Figure 8.1 shows examples of numbers generated from uniform and normal distribution functions.

One can also generate random numbers for any distribution using a *quantile* function for the distribution. This is the inverse of the cumulative distribution function we discussed in chapter 6; instead of identifying the cumulative probabilities for a set of values, the quantile function identifies the values for a set of cumulative probabilities. Using the quantile function, we can generate random numbers from a uniform distribution and then map those numbers into the distribution of interest via its quantile function.

By default, the random number generators in statistical software generate a different set of random numbers every time they are run. However, it is also possible to generate exactly the same set of random numbers, by setting what is called the *random seed* to a specific value. If you were to look at the code that generated these figures (which is available via https://press.princeton.edu/isbn/9780691218441, under Resources), you would see that the random seed is set to a specific value for each chapter in order to make sure that the examples are reproducible.

Using Monte Carlo Simulation

Let's go back to our example of exam finishing times. Say that I administer three quizzes and record the finishing times for each student for each exam, which might look like the distributions presented in figure 8.2.

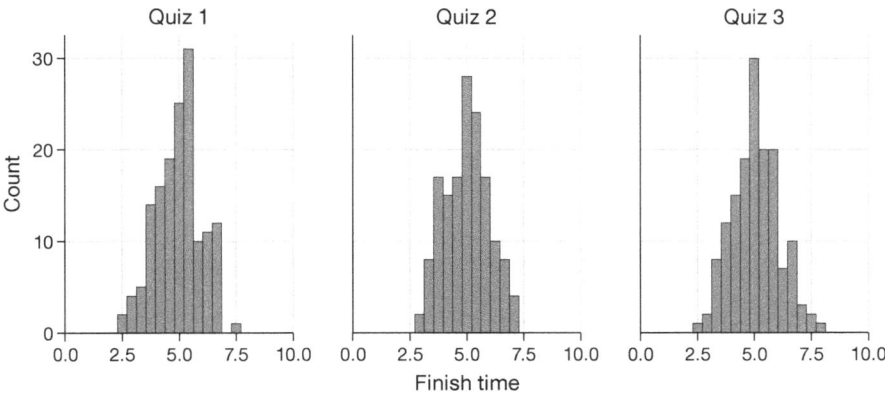

FIGURE 8.2. Simulated finishing time distributions.

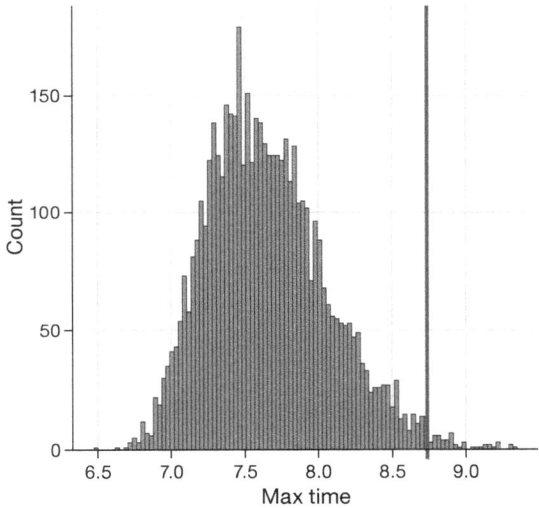

FIGURE 8.3. Distribution of maximum finishing times across simulations. The vertical line marks the ninety-ninth percentile of the finishing time distribution.

What we really want to know is not what the distribution of finishing times looks like, but rather what the distribution of the *longest* finishing time for each quiz looks like. To do this, we can simulate the finishing time for a quiz, using the assumption that the finishing times are distributed normally, as stated above; for each of these simulated quizzes, we then record the longest finishing time. We repeat this simulation a large number of times (5000 should be enough) and record the distribution of finishing times, which is shown in figure 8.3.

This shows that the ninety-ninth percentile of the finishing time distribution falls at 8.74, meaning that if we were to give that much time for the quiz, then everyone should finish 99% of the time. It's always important to remember that our assumptions matter—if they are wrong, then the results of the simulation are useless. In this case, we assumed that the finishing time distribution was normally distributed with a particular mean and standard deviation; if these assumptions are incorrect (and they almost certainly are, since it's rare for elapsed times to be normally distributed), then the true answer could be very different.

Using Simulation for statistics: The Bootstrap

So far we have used simulation to demonstrate statistical principles, but we can also use simulation to answer real statistical questions. In this section, we introduce a concept known as the *bootstrap*, which lets us use simulation to quantify our uncertainty about statistical estimates. Later in the book, we will see other examples of how simulation can often be used to answer statistical questions, especially when theoretical statistical methods are not available or when their assumptions are too difficult to meet.

Computing the Bootstrap

In the previous chapter, we used our knowledge of the sampling distribution of the mean to compute the standard error of the mean. But what if we can't assume that the estimates are normally distributed or we don't know their distribution? The idea of the bootstrap is to use the data themselves to estimate an answer. The name comes from the idea of pulling one self up by one's own bootstraps—expressing the idea that we don't have any external source of leverage, so we have to rely on the data themselves. The bootstrap method was conceived by Bradley Efron of the Stanford Department of Statistics, who is one of the world's most influential statisticians.

The idea behind the bootstrap is that we repeatedly sample from the actual dataset; importantly, we sample *with replacement*, such that the same data point will often end up being represented multiple times within one of the samples. We then compute our statistic of interest on each of the bootstrap samples, and use the distribution of those estimates as our sampling distribution. In a sense, we treat our particular sample as the entire population and then repeatedly sample with replacement to generate our samples for analysis. This makes the assumption that our particular sample is an accurate reflection of the population, which is usually reasonable for larger samples but can break down when samples are smaller.

Let's start by using the bootstrap to estimate the sampling distribution of the mean of adult height in the NHANES dataset, so that we can compare the result to the standard error of the mean (SEM) that we discussed in chapter 7. Figure 8.4 shows that the distribution of means across bootstrap samples is fairly close to the theoretical estimate based on the assumption of normality. We would not usually employ the bootstrap to compute confidence intervals for the mean (since we can generally assume that the normal distribution is appropriate for the sampling distribution of the mean, as long as our sample is large enough), but this example shows how the method gives us roughly the same result as

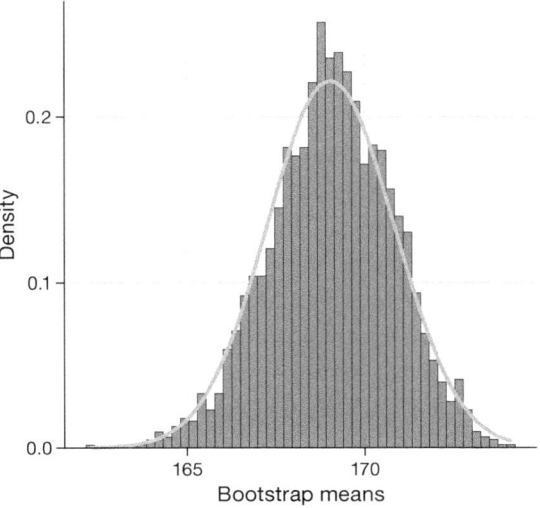

FIGURE 8.4. An example of bootstrapping to compute the standard error of the mean adult height in the NHANES dataset. The histogram shows the distribution of means across bootstrap samples, while the curve shows the normal distribution based on the sample mean and standard error.

the standard method based on the normal distribution. The bootstrap would more often be used to generate standard errors for estimates of other statistics where we don't have a formula for the standard error or where we know or suspect that the normal distribution is not appropriate.

Suggested Reading

- *Computer Age Statistical Inference: Algorithms, Evidence and Data Science*, by Bradley Efron and Trevor Hastie. This book describes the many ways in which computational advances have enabled new statsitical methods, such as the bootstrap. Be forewarned that it is quite mathematically detailed.

Problems

1. Match each of these terms with the appropriate procedure in the following list: bootstrap, Monte Carlo simulation, sampling error, sampling with replacement, sampling without replacement
   - Repeatedly sample from a dataset, compute the statistic on each sample, and combine the samples to estimate a sampling distribution.

- Generate random numbers from a particular distribution, perform a computation, and combine across many repetitions.
- Obtain multiple samples from a population and examine the variability of the statistics computed on each sample.
- Repeatedly sample from a population, putting the sampled individual back into the population so that it can be sampled again.
- Repeatedly sample from a population, setting the sampled individual aside so that they cannot be sampled again.

2. What are the four steps in performing a Monte Carlo simulation?
3. Describe two of the ways in which humans are bad at understanding randomness.
4. What does one need to do in order to regenerate exactly the same sequence of pseudorandom numbers in a statistical programming language?
   - Restart the software and rerun the code.
   - Set the random seed to the same number each time.
   - It is not possible to regenerate exactly the same sequence of pseudorandom numbers twice.
5. What is the primary goal of the bootstrap technique?
   - To generate truly random numbers
   - To estimate the sampling error of a statistic using the data
   - To estimate the mean of a dataset
   - To create a normal distribution

9

Hypothesis Testing

In the first chapter, we discussed the three major goals of statistics: *describe*, *decide*, and *predict*. In this chapter, we introduce the ideas behind the use of statistics to make decisions—in particular, decisions about whether a specific hypothesis is supported by the data.

Learning Objectives

Having read this chapter, you should be able to

- Identify the components of a hypothesis test, including the parameter of interest, the null and alternative hypotheses, and the test statistic.
- Describe the proper interpretations of a p-value as well as common misinterpretations.
- Distinguish between the two types of errors in hypothesis testing and explain the factors that determine them.
- Describe how resampling can be used to compute a p-value.
- Describe the problem of multiple testing and how it can be addressed.
- Describe the main criticisms of null hypothesis statistical testing.

Null Hypothesis Statistical Testing

The specific type of hypothesis testing that we discuss here is known as *null hypothesis statistical testing*, or NHST (for reasons that will become clear). If you pick up almost any scientific or biomedical research publication, you will see NHST being used to test hypotheses, and in their introductory psychology textbook, Gerrig and Zimbardo (2002) refer to NHST as the "backbone of psychological research." Thus, learning how to use and interpret the results from hypothesis testing is essential to understand the results from many fields of research.

It is also important to note, however, that NHST is deeply flawed and that many statisticians and researchers (including myself) think it has been the cause of serious problems in science, which we discuss in chapter 18. For more than 50 years, there have been calls to abandon NHST in favor of other approaches (like those that we discuss in the following chapters):

- "The test of statistical significance in psychological research may be taken as an instance of a kind of essential mindlessness in the conduct of research." (Bakan 1966)
- Hypothesis testing is "a wrongheaded view about what constitutes scientific progress." (Luce 1988)

NHST is also widely misunderstood, largely because it violates our intuitions about how statistical hypothesis testing should work. Let's look at an example to see this.

Null Hypothesis Statistical Testing: An Example

There is great interest in the use of body-worn cameras by police officers, which are thought to reduce the use of force and improve officer behavior. However, in order to establish this, we need experimental evidence, and it has become increasingly common for governments to use randomized controlled trials to test such ideas. A randomized controlled trial of the effectiveness of body-worn cameras was performed by the Washington, DC, city government and the DC Metropolitan Police Department in 2015/2016. Officers were randomly assigned to wear a body-worn camera or not, and their behavior was then tracked over time to determine whether the cameras resulted in less use of force and fewer civilian complaints about officer behavior.

Before we get to the results, how do you think the statistical analysis might work? Let's say we want to specifically test the hypothesis of whether the use of force is decreased by the wearing of cameras. The randomized controlled trial provides us with the data to test the hypothesis—namely, the rates of use of force by officers assigned to either the camera or control groups. The next obvious step is to look at the data and determine whether they provide convincing evidence for or against this hypothesis. That is, What is the likelihood that body-worn cameras reduce the use of force, given the data and everything else we know?

It turns out that this is *not* how null hypothesis testing works. Instead, we first take our hypothesis of interest (i.e., that body-worn cameras reduce use of force) and flip it on its head, creating a *null hypothesis*—in this case, the null hypothesis would be that cameras do not reduce the use of force. Importantly, we then assume that the null hypothesis is true. We then look at the data and determine how likely the data would be if the null hypothesis were true. If the data are sufficiently unlikely under the null hypothesis, then we can reject the null in favor of the *alternative hypothesis*, which is our hypothesis of interest. If there is not sufficient evidence to reject the null, then we say that we retain (or "fail to reject") the null, sticking with our initial assumption that the null is true.

Understanding some of the concepts of NHST, particularly the notorious "p-value," is invariably challenging the first time one encounters them, because they are so counterintuitive. As we will see later, there are other approaches that provide a much more intuitive way to address hypothesis testing (but have their own complexities). However, before we get to those, it's important for you to have a deep understanding of how hypothesis testing works, because it's clearly not going to go away any time soon.

The Process of Null Hypothesis Testing

We can break down the process of null hypothesis statistical testing into a number of steps:

1. Formulate a hypothesis that embodies our prediction (*before seeing the data*).
2. Specify null and alternative hypotheses.
3. Collect some data relevant to the hypothesis.
4. Fit a model to the data that represents the alternative hypothesis and compute a test statistic.
5. Compute the probability of the observed value of that statistic assuming that the null hypothesis is true.
6. Assess the "statistical significance" of the result.

For a hands-on example, let's use the NHANES data to ask the following question: Is physical activity related to blood pressure? In the NHANES dataset, participants were asked whether they engage regularly in moderate- or vigorous-intensity sports, fitness, or recreational activities (stored in the variable *PhysActive*). The researchers also measured blood pressure three times; for our purposes, we will look at the average of the three systolic blood pressure measurements (the higher of the two blood pressure values).

Step 1: Formulate a Hypothesis of Interest

We hypothesize that blood pressure is greater for people who do not engage in physical activity, compared to those who do.

Step 2: Specify the Null and Alternative Hypotheses

For step 2, we need to specify our null hypothesis (which we call H_0) and our alternative hypothesis (which we call H_A). H_0 is the baseline against which we test our hypothesis of interest—that is, what would we expect the data to look like if there were no effect? The null hypothesis always involves some kind of equality ($=$, $\leq$, or $\geq$). H_A describes what we would expect if there actually is an effect. The alternative hypothesis always involves some kind of inequality ($\neq$, $>$, or $<$). Importantly, null hypothesis statistical testing operates under the assumption that the null hypothesis is true unless the data are sufficiently unlikely under the null hypothesis to convince us otherwise.

We also have to decide whether we want to test a *directional* or *nondirectional* hypothesis. A nondirectional hypothesis simply predicts that there will be a difference, without predicting which direction it will go. For the blood pressure/activity example, a nondirectional null hypothesis would be

$$H_0 : BP_{active} = BP_{inactive}$$

Table 9.1. Summary of Blood Pressure Data for Active versus Inactive Individuals

PhysActive	Number	Mean	Standard deviation
No	126	124	21
Yes	124	119	17

and the corresponding nondirectional alternative hypothesis would be

$$H_A : BP_{active} \neq BP_{inactive}$$

A directional hypothesis, on the other hand, predicts which direction the difference would go. For example, we have strong prior knowledge to predict that people who engage in physical activity should have lower blood pressure than those who do not, so we would propose the directional null hypothesis to be

$$H_0 : BP_{active} \geq BP_{inactive}$$

and the directional alternative to be

$$H_A : BP_{active} < BP_{inactive}$$

As we will see later, testing a nondirectional hypothesis is more conservative, so this is generally preferred unless there is a strong a priori reason to hypothesize an effect in a particular direction. Importantly, hypotheses should always be specified (including whether they are directional or not) prior to looking at the data!

Step 3: Collect Some Data

In this case, we will sample 250 individuals from the NHANES dataset. Figure 9.1 shows an example of such a sample, with blood pressure shown separately for active and inactive individuals, and table 9.1 shows summary statistics for each group.

Step 4: Fit a Model to the Data and Compute a Test Statistic

We next want to use the data to compute a statistic that will ultimately let us decide whether the null hypothesis is rejected or not. To do this, the model needs to quantify the amount of evidence in favor of the alternative hypothesis relative to the variability of the statistic. Thus, we can think of the test statistic as providing a measure of the size of the effect compared to the variability in the statistic. In general, this test statistic will have a probability distribution associated with it, because that allows us to determine how likely our observed value of the statistic is under the null hypothesis.

For the blood pressure example, we need a test statistic that allows us to test for a difference between two means, since the hypotheses are stated in terms of the mean blood

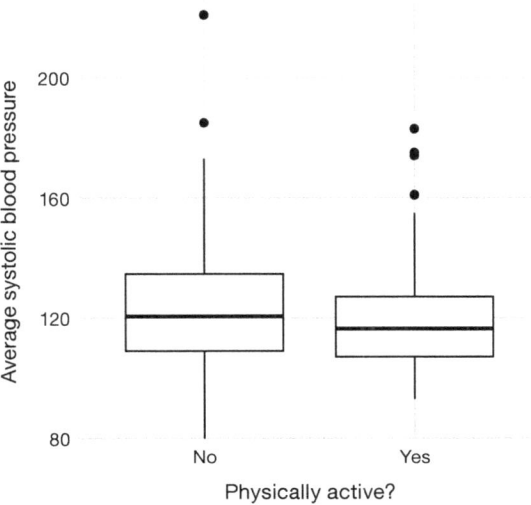

FIGURE 9.1. Box plot of blood pressure data from
a sample of adults from the NHANES dataset,
split by whether they reported engaging in regular
physical activity.

pressure for each group. One statistic that is often used to compare two means is the t statistic, first developed by the statistician William Sealy Gossett, who worked for the Guiness Brewery in Dublin and wrote under the pen name Student—hence, it is often called "Student's t statistic." The t statistic is appropriate for comparing the means of two groups when the sample sizes are relatively small and the population standard deviation is unknown. The t statistic for comparison of two independent groups is computed as

$$t = \frac{\bar{X}_1 - \bar{X}_2}{\sqrt{\frac{S_1^2}{n_1} + \frac{S_2^2}{n_2}}}$$

where $\bar{X}_1$ and $\bar{X}_2$ are the means of the two groups, S_1^2 and S_2^2 are the estimated variances of the groups, and n_1 and n_2 are the sizes of the two groups. Because the variance of a difference between two independent variables is the sum of the variances of each individual variable $(var(A - B) = var(A) + var(B))$, we add the variances for each group divided by their sample sizes in order to compute the standard error of the difference. Thus, one can view the t statistic as a way of quantifying how large the difference between groups is in relation to the sampling variability of the difference between means.

The t statistic is distributed according to a probability distribution known as a t distribution. The t distribution looks quite similar to a normal distribution, but it differs depending on the number of degrees of freedom. When the degrees of freedom are large (say, 1000), then the t distribution looks essentially the same as the normal distribution; but when they are small, then the t distribution has wider tails than the normal (figure 9.2).

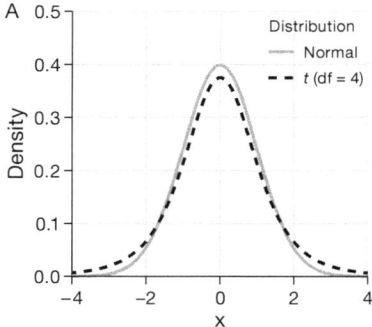 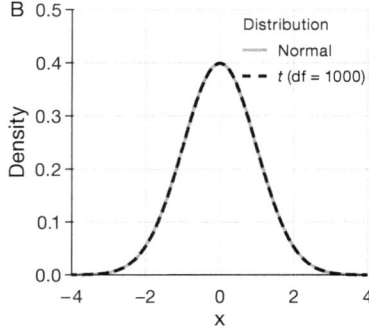

FIGURE 9.2. Each panel shows the *t* distribution (dashed line) overlaid on the normal distribution (solid line). Panel A shows a *t* distribution with 4 degrees of freedom, in which case the distribution is similar but has slightly wider tails. Panel B shows a *t* distribution with 1000 degrees of freedom, in which case it is virtually identical to the normal.

In the simplest case, where the groups are the same size and have equal variance, the degrees of freedom for the *t* test are the number of observations minus two, since we have computed two means and thus given up two degrees of freedom. In this case, it's pretty clear from the box plot that the inactive group is more variable than the active group and the numbers in each group differ, so we need to use a slightly more complex formula for the degrees of freedom, which is often referred to as a *Welch t test*. The formula is

$$degrees\ of\ freedom\ (df) = \frac{\left(\frac{S_1^2}{n_1} + \frac{S_2^2}{n_2}\right)^2}{\frac{\left(S_1^2/n_1\right)^2}{n_1-1} + \frac{\left(S_2^2/n_2\right)^2}{n_2-1}}$$

This will be equal to $n_1 + n_2 - 2$ when the variances and sample sizes are equal, and otherwise will be smaller, in effect imposing a penalty on the test for differences in sample size or variance. For this example, that comes out to 246.41, which is slightly below the value of 248 that one would get by subtracting two from the sample size.

Step 5: Determine the Probability of the Observed
Result Under the Null Hypothesis

This is the step where NHST starts to violate our intuition. Rather than determining the likelihood that the null hypothesis is true given the data, we instead determine the likelihood under the null hypothesis of observing a statistic at least as extreme as one that we have observed—because we started out by assuming that the null hypothesis is true! To do this, we need to know the expected probability distribution for the statistic under the null hypothesis, so that we can ask how likely the result would be under that distribution. Note that when I say "how likely the result would be," what I really mean is "how likely the observed result or one more extreme would be." There are (at least) two reasons that we need to add this caveat. The first is that, when we are talking about continuous values, the probability of any particular value is zero (as you might remember if you've taken a calculus class). More importantly, we are trying to determine how weird our result would be if the null hypothesis were true, and any result that is more extreme would be even more weird, so we want to count all of those weirder possibilities when we compute the probability of our result under the null hypothesis.

We can obtain this "null distribution" either using a theoretical distribution (like the t distribution) or using randomization. Before we move to our blood pressure example, let's start with some simpler examples.

P-VALUES: A VERY SIMPLE EXAMPLE

Let's say that we wish to determine whether a particular coin is biased toward landing on heads. To collect data, we flip the coin 100 times, and let's say we count 70 heads. In this example, $H_0 : P(heads) \leq 0.5$ and $H_A : P(heads) > 0.5$, and our test statistic is simply the number of heads that we counted. The question that we then want to ask is, How likely is it that we would observe 70 or more heads in 100 coin flips if the true probability of heads is 0.5? We can imagine that this might happen very occasionally just by chance, but it doesn't seem very likely. To quantify this probability, we can use the binomial distribution that we encountered in chapter 6:

$$P(X \leq k) = \sum_{i=0}^{k} \binom{N}{k} p^i (1-p)^{(n-i)}$$

This equation tells us the probability of a certain number of heads (k) or fewer, given a particular probability of heads (p) and number of events (N). However, what we really want to know is the probability of a certain number or more, which we can obtain by subtracting from one, based on the rules of probability:

$$P(X \geq k) = 1 - P(X < k)$$

Using the binomial distribution, the probability of 69 or fewer heads given $P(heads) = 0.5$ is 0.999961, so the probability of 70 or more heads is simply one minus that value (0.000039). This computation shows us that the likelihood of getting 70 or more heads if the coin is indeed fair is very small.

Now, what if we didn't have a standard function to tell us the probability of that number of heads? We could instead determine it by simulation—we repeatedly flip a coin 100 times using a true probability of 0.5 and then compute the distribution of the number of heads across those simulation runs. Figure 9.3 shows the result from this simulation. Here we can see that the probability computed via simulation (0.00002) is very close to the theoretical probability (0.00004).

COMPUTING P-VALUES USING THE t DISTRIBUTION

Now let's compute a p-value for our blood pressure example using the t distribution. First, we compute the t statistic using the values from our sample that we calculated above, where we find that $t = 1.89$. The question that we then want to ask is, What is the

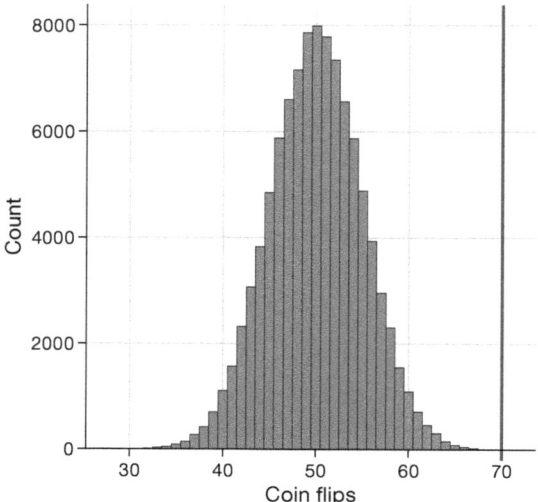

FIGURE 9.3. Distribution of the number of heads (out of 100 flips) across 100,000 simulated runs, with the observed value of 70 flips represented by the vertical line.

likelihood that we would find a t statistic of this size, if the true difference between groups is zero or less (i.e., the directional null hypothesis)?

We can use the t distribution to determine this probability. Earlier we noted that the appropriate degrees of freedom (after correcting for differences in variance and sample size) was $t = 246.41$. We can use a function from our statistical software to determine the probability of finding a value of the t statistic greater than or equal to our observed value. We find that $P(t > 1.89, \; df = 246.41) = 0.029857$, which tells us that our observed t statistic value of 1.89 is relatively unlikely if the null hypothesis really is true.

In this case, we used a directional hypothesis, so we only had to look at one end of the null distribution. If we wanted to test a nondirectional hypothesis, then we would need to be able to identify how unexpected the size of the effect is, regardless of its direction. In the context of the t test, this means that we need to know how likely it is that the statistic would be as extreme in either the positive or negative direction. To do this, we multiply the observed t value by -1, since the t distribution is centered around zero, and then add together the two tail probabilities to get a *two-tailed* p-value: $P(t > 1.89 \text{ or } t < -1.89, df = 246.41) = 0.059713$. Here we see that the p-value for the two-tailed test is twice as large as that for the one-tailed test, which reflects the fact that an extreme value is less surprising since it could have occurred in either direction.

How do you choose whether to use a one-tailed versus a two-tailed test? The two-tailed test is always going to be more conservative, so it's always a good bet to use that one, unless you had a very strong prior reason for using a one-tailed test. In that case, you should have written down the hypothesis before you ever looked at the data. In chapter 18, we discuss

Table 9.2. Squatting Data for the
Two Groups

Group	Squat	Shuffled squat
FB	265	125
FB	310	230
FB	335	125
FB	230	315
FB	315	115
XC	155	335
XC	125	155
XC	125	125
XC	125	265
XC	115	310

the idea of preregistration of hypotheses, which formalizes the idea of writing down your hypotheses before you ever see the actual data. You should *never* make a decision about how to perform a hypothesis test once you have looked at the data, as this can introduce serious bias into the results.

COMPUTING P-VALUES USING RANDOMIZATION

So far we have seen how we can use the t distribution to compute the probability of the data under the null hypothesis, but we can also do this using simulation. The basic idea is that we generate simulated data like those that we would expect under the null hypothesis, and then ask how extreme the observed data are in comparison to those simulated data. The key question is, How can we generate data for which the null hypothesis is true? The general answer is that we can randomly rearrange the data in a particular way that makes the data look like they would if the null were really true. This is similar to the idea of bootstrapping, in the sense that it uses our own data to come up with an answer, but it does it in a different way.

RANDOMIZATION: A SIMPLE EXAMPLE

Let's start with a simple example. Let's say that we want to compare the mean squatting ability of football players with cross-country runners, with $H_0 : \mu_{FB} \leq \mu_{XC}$ and $H_A : \mu_{FB} > \mu_{XC}$. We measure the maximum squatting ability of five football players and five cross-country runners (which we generate randomly, assuming that $\mu_{FB} = 300$, $\mu_{XC} = 140$, and $\sigma = 30$). The data are shown in table 9.2.

From the plot in Panel A of figure 9.4, it's clear that there is a large difference between the two groups. We can do a standard t test to test our hypothesis; for this example, we use the `t.test()` command in R, which gives the following result:

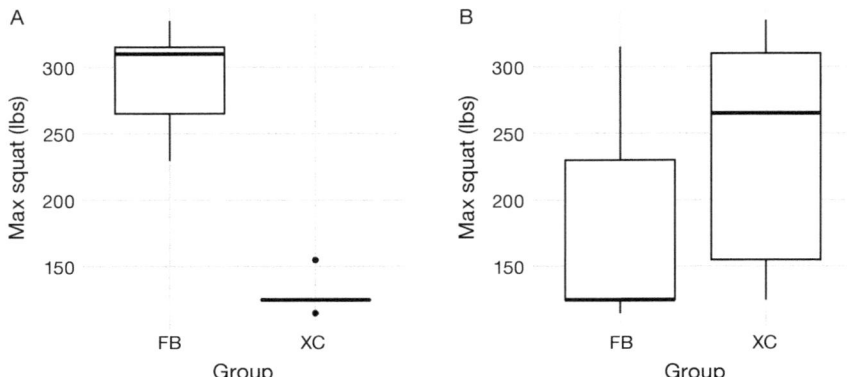

FIGURE 9.4. (A) Box plots of simulated squatting ability for football players and cross-country runners. (B) Box plots for subjects assigned to each group after scrambling group labels.

```
##
##   Welch Two-Sample t-test
##
## data:   squat by group
## t = 8, df = 5, p-value = 0.0002
## alternative hypothesis: true difference in means between
## group FB and group XC is greater than 0
## sample estimates:
## mean in group FB    mean in group XC
## 291                 129
```

If we look at the p-value reported here, we see that the likelihood of such a difference under the null hypothesis is very small, using the t distribution to define the null.

Now let's see how we could answer the same question using randomization. The basic idea is that, if the null hypothesis of no difference between groups is true, then it shouldn't matter which group the individuals come from (football players versus cross-country runners). Thus, to create data that are like our actual data but that also conform to the null hypothesis, we can randomly relabel individuals in the dataset in terms of their group membership (by randomly shuffling the labels) and then recompute the difference between the groups. The results of such a shuffle are shown in the last column of table 9.2, and the box plots of the resulting data are in panel B of figure 9.4.

After scrambling the data, we see that the two groups are now much more similar, and in fact the cross-country group now has a slightly higher mean. Now let's do that 10,000 times and store the t statistic for each iteration; if you are doing this on your own computer, it will take a moment to complete. Figure 9.5 shows the histogram of the t values across all the random shuffles. As expected under the null hypothesis, this distribution is centered at zero (the mean of the distribution is 0.007). From the figure, we can also see that

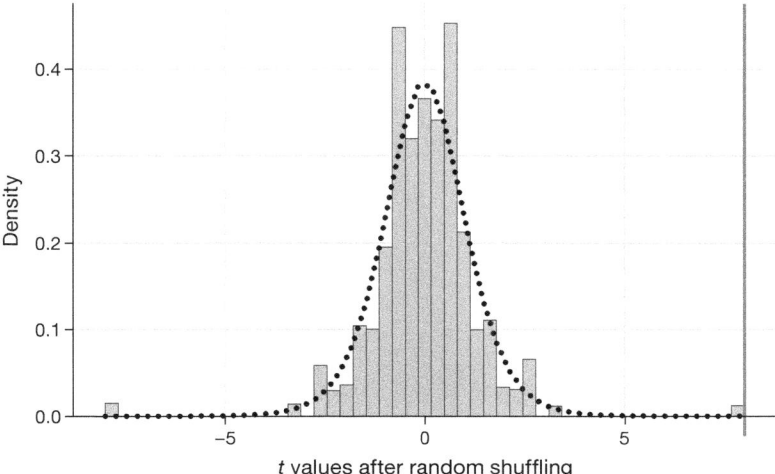

FIGURE 9.5. Histogram of t values for the difference in means between the football and cross-country groups after randomly shuffling group membership. The vertical line to the right denotes the actual difference observed between the two groups, and the dotted line shows the theoretical t distribution for this analysis.

the distribution of t values after shuffling roughly follows the theoretical t distribution under the null hypothesis (with mean $= 0$), showing that randomization worked to generate null data. We can compute the p-value from the randomized data by measuring how many of the shuffled values are at least as extreme as the observed value: $P(t > 8.01, \ df = 8)$ using randomization $= 0.00410$. This p-value is very similar to the p-value that we obtained using the t distribution, and both are quite extreme, suggesting that the observed data are very unlikely to have arisen if the null hypothesis is true—and in this case we *know* that it's not true, because we generated the data.

RANDOMIZATION: BLOOD PRESSURE/ACTIVITY EXAMPLE

Now let's use randomization to compute the p-value for the blood pressure/activity example. In this case, we will randomly shuffle the *PhysActive* variable and compute the difference between groups after each shuffle, and then compare our observed t statistic to the distribution of t statistics from the shuffled datasets. Figure 9.6 shows the distribution of t values from the shuffled samples, and we can also compute the probability of finding a value as large or larger than the observed value. The p-value obtained from randomization (0.028) is very similar to the one obtained using the t distribution (0.030). The advantage of the randomization test is that it doesn't require that we assume that the data from each of the groups are normally distributed, though the t test is generally quite robust to violations of that assumption. In addition, the randomization test can allow us to compute p-values for statistics when we don't have a theoretical distribution, like we do for the t test.

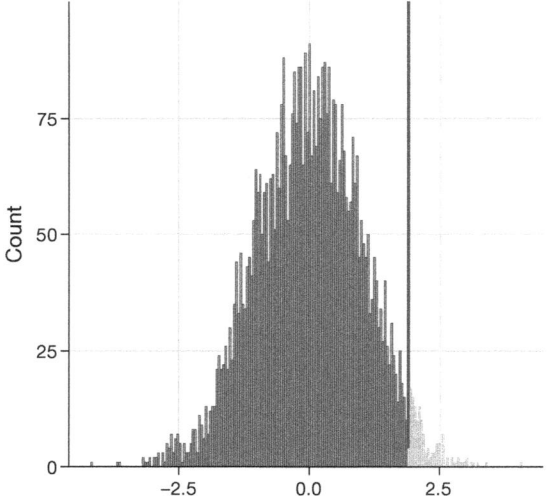

FIGURE 9.6. Histogram of *t* statistics after shuffling of group labels, with the observed value of the *t* statistic shown by the vertical line and values at least as extreme as the observed value shown in light gray.

When we use the randomization test, we do have to make one main assumption, which we refer to as *exchangeability*. This means that all of the observations are distributed in the same way, such that we can interchange them without changing the overall distribution. The main place where this can break down is when there are related observations in the data; for example, if we had data from individuals in four different families, then we couldn't assume that individuals were exchangeable, because siblings would be closer to each other than they are to individuals from other families. In general, if the data were obtained by random sampling, then the assumption of exchangeability across observations should hold.

Step 6: Assess the "Statistical Significance" of the Result

The next step is to determine whether the p-value that results from the previous step is small enough that we are willing to reject the null hypothesis and conclude instead that the alternative is true. How much evidence do we require? This is one of the most controversial questions in statistics, in part because it requires a subjective judgment—there is no "correct" answer.

Historically, the most common answer to this question has been that we should reject the null hypothesis if the p-value is less than 0.05. This comes from the writings of Ronald Fisher, who has been referred to as "the single most important figure in 20th century statistics" (Efron, 1998, p. 95):

If P is between .1 and .9 there is certainly no reason to suspect the hypothesis tested. If it is below .02 it is strongly indicated that the hypothesis fails to account for the whole of the facts. We shall not often be astray if we draw a conventional line at .05 . . . it is convenient to draw the line at about the level at which we can say: Either there is something in the treatment, or a coincidence has occurred such as does not occur more than once in twenty trials. (Fisher 1925, p. 82)

However, Fisher never intended $p < 0.05$ to be a fixed rule:

No scientific worker has a fixed level of significance at which from year to year, and in all circumstances, he rejects hypotheses; he rather gives his mind to each particular case in the light of his evidence and his ideas. (Fisher 1956, p. 41)

Instead, it is likely that $p < .05$ became a ritual due to the reliance on tables of p-values that were used before computing made it easy to generate p-values for arbitrary values of a statistic. All the tables had an entry for 0.05, making it easy to determine whether one's statistic exceeded the value needed to reach that level of significance.

The choice of statistical thresholds remains deeply controversial; recently, it has been proposed that the default threshold be changed from .05 to .005, making it substantially more stringent and thus more difficult to reject the null hypothesis (Benjamin et al. 2018). In large part, this move is due to growing concerns that the evidence obtained from a significant result at $p < .05$ is relatively weak; we return to this in our discussion of reproducibility in chapter 18.

HYPOTHESIS TESTING AS DECISION MAKING: THE NEYMAN-PEARSON APPROACH

Whereas Fisher thought that the p-value could provide evidence regarding a specific hypothesis, the statisticians Jerzy Neyman and Egon Pearson disagreed vehemently. Instead, they proposed that we think of hypothesis testing in terms of its error rate in the long run:

No test based upon a theory of probability can by itself provide any valuable evidence of the truth or falsehood of a hypothesis. But we may look at the purpose of tests from another viewpoint. Without hoping to know whether each separate hypothesis is true or false, we may search for rules to govern our behaviour with regard to them, in following which we insure that, in the long run of experience, we shall not often be wrong. (Neyman and Pearson 1933, p. 291)

That is, we can't know which specific decisions are right or wrong, but if we follow the rules, we can at least know how often our decisions will be wrong in the long run.

To understand the decision-making framework that Neyman and Pearson developed, we first need to discuss statistical decision making in terms of the kinds of outcomes that can occur. There are two possible states of reality (H_0 is true or H_0 is false) and two possible decisions (reject H_0 or retain H_0). There are two ways in which we can make a correct decision:

- We can reject H_0 when it is false (in the language of signal detection theory, we call this a *hit*).

- We can retain H_0 when it is true (somewhat confusingly in this context, this is called a *correct rejection*).

There are also two kinds of errors we can make:

- We can reject H_0 when it is actually true (we call this a *false alarm*, or *Type I error*).
- We can retain H_0 when it is actually false (we call this a *miss*, or *Type II error*).

Neyman and Pearson coined two terms to describe the probability of these two types of errors in the long run:

- $P(Type\ I\ error) = \alpha$
- $P(Type\ II\ error) = \beta$

That is, if we set α to .05, then in the long run we should make a Type I error 5% of the time. Whereas it's common to set α as .05, the standard value for an acceptable level of β is .2—that is, we are willing to accept that 20% of the time we will fail to detect a true effect when it exists. We return to this topic in chapter 10 when we discuss statistical power, which is the complement of Type II error.

What Does a Significant Result Mean?

There is a great deal of confusion about what p-values actually mean (Gigerenzer 2004). Let's say that we do an experiment comparing the means between conditions and we find a difference with a p-value of .01. There are a number of possible interpretations that we might entertain.

Does it mean that the probability of the null hypothesis being true is .01? No. Remember that in null hypothesis testing, the p-value is the probability of the data given the null hypothesis ($P(data|H_0)$). It does not warrant conclusions about the probability of the null hypothesis given the data ($P(H_0|data)$). We return to this question when we discuss Bayesian inference in chapter 11, as Bayes' theorem lets us invert the conditional probability in a way that allows us to determine the probability of a hypothesis given the data.

Does it mean that the probability that you are making the wrong decision is .01? No. This would be $P(H_0|data)$, but remember as above that p-values are probabilities of data under H_0, not probabilities of hypotheses.

Does it mean that if you ran the study again, you would obtain the same result 99% of the time? No. The p-value is a statement about the likelihood of a particular dataset under the null hypothesis; it does not allow us to make inferences about the likelihood of future events, such as replication.

Does it mean that you have found a practically important effect? No. There is an essential distinction between *statistical significance* and *practical significance*. As an example, let's say that we perform a randomized controlled trial to examine the effect of a particular exercise intervention on blood pressure and we find a statistically significant effect at

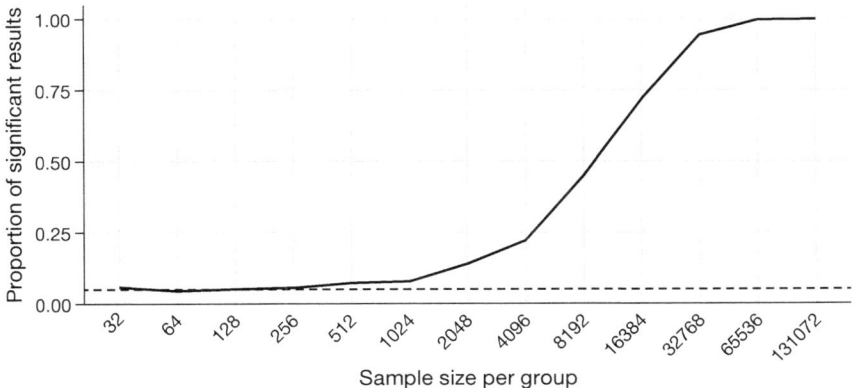

FIGURE 9.7. The proportion of significant results for a very small change (one-half point in systolic blood pressure) as a function of sample size.

p < .05. What this doesn't tell us is the degree of reduction in blood pressure, which we refer to as the *effect size* (to be discussed in more detail in chapter 10). If we think about a study of blood pressure, we probably don't think that a reduction of one-half of one point (which is below the precision of most blood pressure measurement devices) is practically significant. Let's look at our ability to detect a significant difference of half a point as the sample size increases.

Figure 9.7 shows how the proportion of significant results increases as the sample size increases, such that with a very large sample size (about 65,000 total subjects), we will find a significant result in more than 90% of studies when there is a one-half-point reduction in blood pressure due to the intervention. While these are statistically significant, most physicians would not consider this amount of blood pressure reduction to be practically or clinically significant. We explore this relationship in more detail when we return to the concept of statistical power in chapter 10, but it should already be clear from this example that statistical significance is not necessarily indicative of practical significance.

NHST in a Modern Context: Multiple Testing

So far we have discussed examples where we are interested in testing a single statistical hypothesis; this is consistent with traditional science, which often measured only a few variables at a time. However, in modern science we can often measure millions of variables per individual. For example, in genetic studies that quantify the entire genome, there may be many millions of genetic variants measured per individual, and in the brain imaging research that my group does, we often collect data from more than 100,000 locations in the brain at once. When standard hypothesis testing is applied in these contexts, bad things can happen unless we take appropriate care.

Let's look at an example to see how this might work. There is great interest in understanding the genetic factors that can predispose individuals to major mental illnesses, such

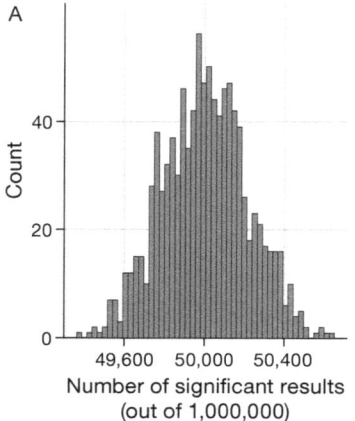
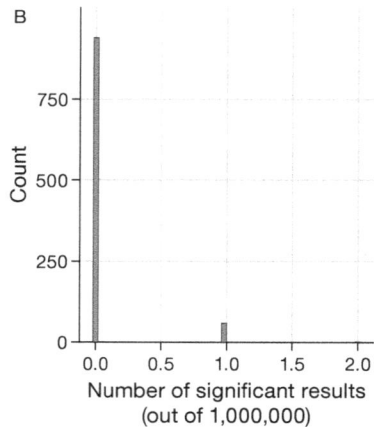

FIGURE 9.8. (A) A histogram of the number of significant results in each set of one million statistical tests, when there is in fact no true effect. (B) A histogram of the number of significant results across all simulation runs after applying the Bonferroni correction for multiple tests.

as schizophrenia, because we know that about 80% of the variation between individuals in the presence of schizophrenia is related to genetic differences. The Human Genome Project and the ensuing revolution in genome science has provided tools to examine the many ways in which humans differ from one another in their genomes. One approach that has been used in recent years is known as a *genome-wide association study* (GWAS), in which the genome of each individual is characterized at one million or more places to determine which letters of the genetic code each person has at each location, focusing on locations where humans frequently tend to differ. After the letters of the genetic code have been determined, the researchers perform a statistical test at each location in the genome to determine whether people diagnosed with schizoprenia are more or less likely to have one specific version of the genetic sequence at that location.

Let's imagine what would happen if the researchers simply asked whether the test was significant at $p < .05$ at each location, when in fact there is no true effect at any of the locations. To do this, we generate a large number of simulated t values from a null distribution and ask how many of them are significant at $p < .05$. Let's do this many times, and each time count up how many of the tests come out as significant (figure 9.8, panel A).

This shows that about 5% of all the tests are significant in each run, meaning that if we were to use $p < .05$ as our threshold for statistical significance, then even if there were no truly significant relationships present, we would still "find" about 50,000 genes that were seemingly significant in each study (the expected number of significant results is simply $n * \alpha$). So while we controlled for the error per test, we didn't control the error rate across our entire *family* of tests (known as the *familywise error*), which is what we really want to control if we are going to be looking at the results from a large number of tests. Using $p < .05$, our familywise error rate in the above example is one—that is, we are pretty much guaranteed to make at least one error in any particular study.

A simple way to control for the familywise error is to divide the alpha level (α) by the number of tests; this is known as the *Bonferroni* correction, named after the Italian statistician Carlo Bonferroni. Using the data from our example above, we see in panel B of figure 9.8 that only about 5% of studies show any significant results using the corrected alpha level of 0.000005 instead of the nominal level of .05. We have effectively controlled the familywise error, such that the probability of making *any* errors in our study is controlled at right around .05.

Suggested Reading

- *The Cult of Statistical Significance: How the Standard Error Costs Us Jobs, Justice, and Lives,* by Stephen Ziliak and Deirdre McCloskey. This book lays out the case for how null hypothesis testing has led to numerous problematic outcomes in the real world.

Problems

1. Describe the six steps in performing null hypothesis testing.
2. The null hypothesis always involves some sort of _____, while the alternative hypothesis always involves some sort of _____. Options: equality, inequality
3. A researcher wishes to test the hypothesis that eating broccoli reduces blood pressure. She randomly assigns one group of participants to eat broccoli four times a week for a month and another group to continue eating their normal diet. She measures blood pressure before and after the intervention and computes the difference between blood pressure before and after the diet (i.e., *difference = BP before diet − BP after diet*).
   - The null hypothesis is that the difference score will be _____.
   - The alternative hypothesis is that the difference will be _____.
4. Which of the following is the most accurate description of the meaning of a p-value in null hypothesis testing?
   - The probability of a statistic at least as extreme as the observed value when the null hypothesis is true
   - The probability that the alternative hypothesis is true given the observed statistic
   - The probability that one would find a value at least as extreme as the observed value if the experiment were run again
   - The probability that the null hypothesis is true given the observed statistic
5. Describe the rationale for using randomization to perform null hypothesis testing.
6. A researcher collects a dataset and computes a t statistic comparing two groups (of size 20 each), which comes out to 2.4. He then performs 10 randomizations of the group label and obtains the following t statistics for the 10 resamples: -0.16, 0.06, 1.07, 2.51, -0.77, -1.27, 1.22, -0.08, 0.49, 0.74.
   - Based on these resamples, what is the p-value for observed value of 2.4?
   - How does this p-value relate to the one obtained using the t distribution?

7. Match each concept in the following list to its definition: Type I error, Type II error, alpha, beta, power.
   - Rejecting the null hypothesis when it is in fact true
   - Failing to reject the null hypothesis when it is in fact false
   - Probability of a Type I error
   - Probability of a Type II error
   - Probability of not making a Type II error

8. A researcher wishes to measure whether the consumption of broccoli is related to income. She obtains data for 200 million people and performs an analysis using a t test that finds a significant p-value of .01. However, the difference in income between people who eat broccoli and people who do not is 4 cents on average. What concept is this most clearly an example of?
   - Practical significance
   - Multiple comparisons
   - Bayesian analysis
   - Randomization testing

9. Which of the following is a true statement about the t distribution? Choose all that apply.
   - The t distribution has wider tails than the normal distribution.
   - The t and normal distributions become more different from one another as the sample size increases.
   - The tails of the t distribution become wider as the sample size increases.
   - Both the normal and t distributions change their shape as the sample size increases.

10. The p-value obtained using a two-tailed test is (larger/smaller) than the one obtained using a one-tailed test.

11. Which of the following is a true statement regarding a p-value of .01? Choose all that apply.
    - There is a 1% chance that the null hypothesis is true.
    - There is a 1% chance that, if the experiment were run again, it would fail to find a significant result.
    - There is a 1% chance that one would find a statistical value at least as extreme as the one observed, if the null hypothesis were true.
    - The alternative hypothesis has been proved true.
    - The result is both statistically and practically significant.

12. Why is it important to control for familywise error when performing multiple hypothesis tests?

13. A researcher performs four hypothesis tests, which result in p-values of .03, .01, .21, and .04. Apply the Bonferroni correction to determine which of these p-values remains significant after correction for multiple tests.

10

Quantifying Effects and Designing Studies

In the previous chapter, we discussed how we can use data to test hypotheses. Those methods provided a binary answer: we either reject or retain the null hypothesis. However, this kind of decision overlooks a couple of important questions. First, we would like to know how much uncertainty we have about the answer (regardless of which way it goes). In addition, sometimes we don't have a clear null hypothesis, so we would like to see what range of estimates are consistent with the data. Second, we would like to know how large the effect actually is, since, as we saw in the blood pressure example in the previous chapter, a statistically significant effect is not necessarily a practically important effect.

In this chapter, we discuss methods to address these two questions: *confidence intervals* to provide a measure of our uncertainty about our estimates, and *effect sizes* to provide a standardized way to understand how large the effects are. We also discuss the concept of *statistical power*, which tells us how likely we are to actually find any true effects that exist.

Learning Objectives

Having read this chapter, you should be able to

- Describe the proper interpretation of a confidence interval and compute a confidence interval for the mean of a given dataset.
- Define the concept of effect size and compute the effect size for a given test.
- Describe the concept of statistical power and why it is important for research.

Confidence Intervals

So far in the book we have focused on estimating a single value for our statistic. For example, let's say we want to estimate the mean weight of adults in the NHANES dataset, so we take a sample from the dataset and estimate the mean. In this sample, the mean weight is 79.92 kilograms. We refer to this as a *point estimate* since it provides us with a single number to describe our estimate of the population parameter. However, we know from our earlier discussion of sampling error that there is some uncertainty about this estimate, which is described by the standard error. You should also remember that the standard error is determined by two components: the population standard deviation (which is the numerator) and the square root of the sample size (which is the denominator). The population standard deviation is a generally unknown but fixed parameter that is not under our control, whereas the sample size *is* under our control. Thus, we can decrease our uncertainty about the estimate by increasing our sample size—up to the limit of the entire population size, at which point there is no uncertainty at all because we can just calculate the population parameter directly from the data of the entire population.

We would often like to have a way to more directly describe our uncertainty about a statistical estimate, which we can accomplish using a confidence interval. Most people are familiar with confidence intervals through the idea of a "margin of error" for political polls. These polls usually try to provide an answer that is accurate within $+/-3\%$. For example, when a candidate is estimated to win an election by 9 percentage points with a margin of error of 3, the percentage by which they will win is estimated to fall within 6–12 percentage points. In statistics, we refer to this as a *confidence interval*, which provides a range of values for our parameter estimate that are consistent with our sample data, rather than just giving us a single estimate based on the data. The wider the confidence interval, the more uncertain we are about our parameter estimate.

Confidence intervals are notoriously confusing, primarily because they don't mean what we might intuitively think they mean. If I tell you that I have computed a 95% confidence interval for my statistic, then it would seem natural to think that we can have 95% confidence that the true parameter value falls within this interval. However, as we have seen throughout the book, concepts in statistics often don't mean what we think they should mean. In the case of confidence intervals, we can't interpret them in this way because the population parameter has a fixed value—it either is or isn't in the interval, so it doesn't make sense to talk about the probability of that occurring. Jerzy Neyman, the inventor of the confidence interval, put it this way:

> The parameter is an unknown constant and no probability statement concerning its value may be made. (Neyman 1937, p. 349)

Instead, we have to view the confidence interval procedure from the same standpoint that we viewed hypothesis testing: as a procedure that in the long run will allow us to make correct statements with a particular probability. Thus, the proper interpretation of the 95% confidence interval is that it is an interval that will contain the true population mean 95% of the time, and in fact we can confirm that using simulation, as you will see below.

The confidence interval for the mean is computed as

$$CI = point\ estimate \pm critical\ value * standard\ error$$

where the critical value is determined by the sampling distribution of the estimate. The important question, then, is how we obtain our estimate for that sampling distribution.

Confidence Intervals Using the Normal Distribution

If we know the population standard deviation, then we can use the normal distribution to compute a confidence interval. We usually don't, but for our example of the NHANES dataset we do, since we are treating the entire dataset as the population (it's 21.3 for weight).

Let's say that we want to compute a 95% confidence interval for the mean. The critical value would then be the values of the standard normal distribution that capture 95% of the distribution; these are simply the 2.5 percentile and the 97.5 percentile of the distribution, which we can compute using our statistical software and go ± 1.96. Thus, the confidence interval for the mean $(\bar{X})$ is

$$CI = \bar{X} \pm 1.96 * SE$$

Using the estimated mean from our sample (79.92) and the known population standard deviation, we can compute the confidence interval of $[77.28, 82.56]$.

Confidence Intervals Using the t Distribution

As stated above, if we knew the population standard deviation, then we could use the normal distribution to compute our confidence intervals. However, in general we don't—in which case the t distribution is more appropriate as a sampling distribution. Remember that the t distribution is slightly broader than the normal distribution, especially for smaller samples, which means that the confidence intervals will be slightly wider than they would if we were using the normal distribution. This incorporates the extra uncertainty that arises when we estimate parameters based on small samples.

We can compute the 95% confidence interval in a way similar to the normal distribution example above, but the critical value is determined by the 2.5 percentile and the 97.5 percentile of the t distribution with the appropriate degrees of freedom. Thus, the confidence interval for the mean $(\bar{X})$ is

$$CI = \bar{X} \pm t_{crit} * SE$$

where t_{crit} is the critical t value. For the NHANES weight example (with sample size of 250), the confidence interval would be $79.92 +/- 1.97 * 1.41 [77.15 - 82.69]$.

Remember that this doesn't tell us anything about the probability of the true population value falling within this interval, since it is a fixed parameter (which we know is 81.77 because we have the entire population in this case) and it either does or does not fall within this specific interval (in this case, it does). Instead, it tells us that, in the long run, if we compute the confidence interval using this procedure, 95% of the time that confidence interval will capture the true population parameter.

We can see this using the NHANES data as our population; in this case, we know the true value of the population parameter, so we can see how often the confidence interval ends up capturing that value across many different samples. Figure 10.1 shows

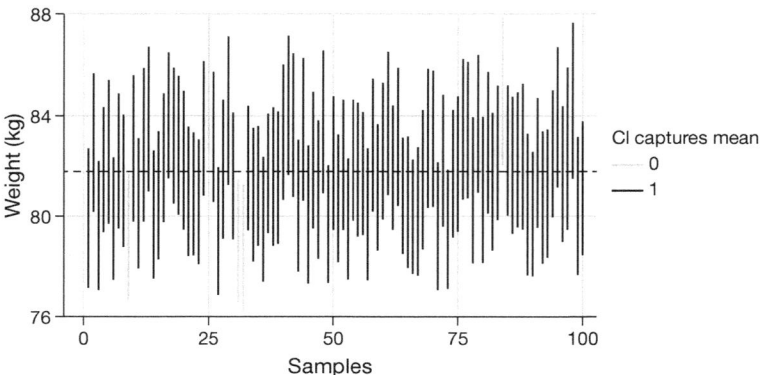

FIGURE 10.1. Samples were repeatedly taken from the NHANES dataset, and the 95% confidence interval of the mean was computed for each sample. Intervals shown in light gray did not capture the true population mean (shown as the dashed line).

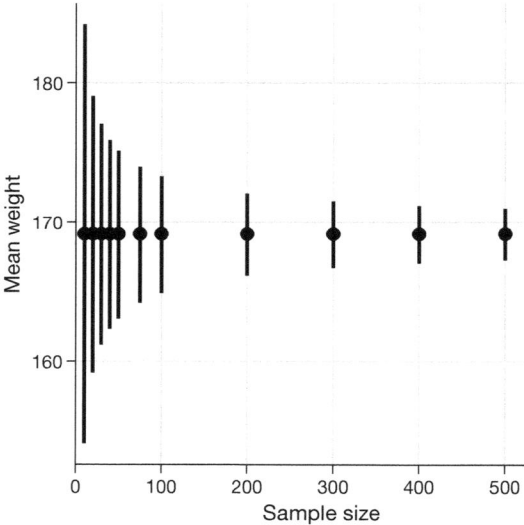

FIGURE 10.2. An example of the effect of sample size on the width of the confidence interval for the mean.

the confidence intervals for estimated mean weight computed for 100 samples from the NHANES dataset. Of these, 95 captured the true population mean weight, showing that the confidence interval procedure performs as it should.

Confidence Intervals and Sample Size

Because the standard error decreases with sample size, the confidence interval should get narrower as the sample size increases, providing progressively tighter bounds on our estimate. Figure 10.2 shows an example of how the confidence interval would change as a function of sample size for the weight example. From the figure, it's evident that the confidence interval becomes increasingly tighter as the sample size increases; but increasing samples provide diminishing returns, consistent with the fact that the denominator of the confidence interval term is proportional to the square root of the sample size.

Computing Confidence Intervals Using the Bootstrap

In some cases, we can't assume normality or we don't know the sampling distribution of the statistic. In these cases, we can use the bootstrap (which we introduced in chapter 8). As a reminder, the bootstrap involves repeatedly resampling the data *with replacement* and then using the distribution of the statistic computed on those samples as a surrogate for the sampling distribution of the statistic. These are the results when we use the built-in bootstrapping function in R to compute the confidence interval for weight in our NHANES sample:

```
## BOOTSTRAP CONFIDENCE INTERVAL CALCULATIONS
## Based on 1000 bootstrap replicates

## Intervals :
## Level      Percentile
## 95%          (78, 84 )
## Calculations and Intervals on Original Scale
```

These values are fairly close to the values obtained using the *t* distribution above, though not exactly the same.

Relation of Confidence Intervals to Hypothesis Tests

There is a close relationship between confidence intervals and hypothesis tests. In particular, if the confidence interval does not include the null hypothesis, then the associated statistical test would be statistically significant. For example, if you are testing whether the mean of a sample is different from zero with $\alpha = 0.05$, you could simply check to see whether zero is contained within the 95% confidence interval for the mean.

Things get trickier if we want to compare the means of two conditions (Schenker and Gentleman 2001). There are a couple of situations that are clear. First, if each mean is contained within the confidence interval for the other mean, then there is definitely no significant difference at the chosen confidence level. Second, if there is no overlap between the confidence intervals, then there is certainly a significant difference at the chosen level; in fact, this test is substantially *conservative*, such that the actual error rate will be lower than the chosen level. But what about the case where the confidence intervals overlap one another but don't contain the means for the other group? In this case, the answer depends on the relative variability of the two variables, and there is no general answer. However, one should in general avoid using the "eyeball test" for overlapping confidence intervals.

Effect Sizes

Statistical significance is the least interesting thing about the results. You should describe the results in terms of measures of magnitude—not just, does a treatment affect people, but how much does it affect them. (Gene Glass, quoted in Sullivan and Feinn 2012)

In the previous chapter, we discussed the idea that statistical significance may not necessarily reflect practical significance. In order to discuss practical significance, we need a standard way to describe the size of an effect in terms of the actual data, which we refer to as an *effect size*. In this section, we introduce the concept and discuss various ways that effect sizes can be calculated.

An effect size is a standardized measurement that compares the size of some statistical effect to a reference quantity, such as the variability of the data. In some fields of science and engineering, this idea is referred to as a *signal-to-noise ratio*. There are many different ways that the effect size can be quantified, which depend on the nature of the data.

Cohen's d

One of the most common measures of effect size is known as *Cohen's d,* named after the statistician Jacob Cohen (who is most famous for his 1994 paper titled "The Earth Is Round (p < .05)"). It is used to quantify the difference between two means, in terms of their standard deviation:

$$d = \frac{\bar{X}_1 - \bar{X}_2}{s}$$

Here $\bar{X}_1$ and $\bar{X}_2$ are the means of the two groups, and s is the pooled standard deviation (which is a combination of the standard deviations for the two samples, weighted by their sample sizes), defined as

$$s = \sqrt{\frac{(n_1 - 1)s_1^2 + (n_2 - 1)s_2^2}{n_1 + n_2 - 2}}$$

where n_1 and n_2 are the sample sizes and s_1^2 and s_2^2 are the standard deviations for the two groups, respectively. Note that this is very similar in spirit to the t statistic—the main difference is that the denominator in the t statistic is based on the standard error of the mean, whereas the denominator in Cohen's d is based on the standard deviation of the data. This means that, while the t statistic grows as the sample size gets larger, the value of Cohen's d remains the same.

There is a commonly used scale for interpreting the size of an effect in terms of Cohen's d, shown in table 10.1. It can be useful to look at some commonly understood effects to help understand these interpretations. For example, the effect size for gender differences in adult height ($d = 2.05$) is very large by reference to our table above. We can also see this

Table 10.1. Interpretation of Cohen's d

d	Interpretation
0.0—0.2	Negligible
0.2—0.5	Small
0.5—0.8	Medium
0.8—	Large

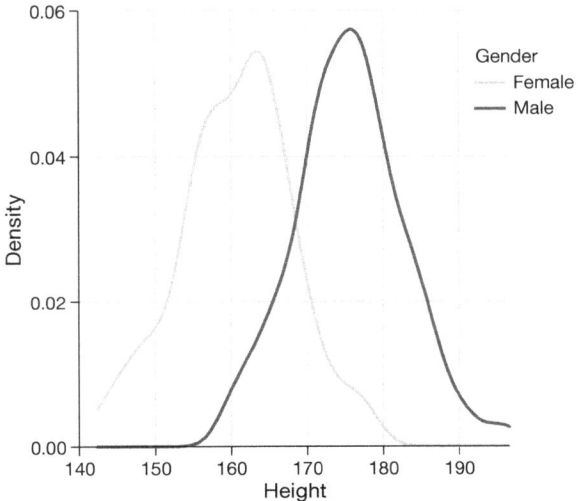

FIGURE 10.3. Smoothed histogram plots for male and female heights in the NHANES dataset, showing clearly distinct but also clearly overlapping distributions.

by looking at the distributions of male and female heights in a sample from the NHANES dataset. Figure 10.3 shows that the two distributions are quite well separated though still overlapping, highlighting the fact that even when there is a very large effect size for the difference between two groups, there will be individuals from each group that are more like the other group.

It is also worth noting that we rarely encounter effects of this magnitude in science, in part because they are such obvious effects that we don't need scientific research to find them. As we will see in chapter 18 on reproducibility, very large reported effects in scientific research often reflect the use of questionable research practices rather than truly huge effects in nature. It is also important to point out that, even for such a huge effect, the two distributions still overlap—there will be some females who are taller than the average male and vice versa. For most interesting scientific effects, the degree of overlap is much greater, so we shouldn't immediately jump to strong conclusions about individuals from different populations based on even a large effect size.

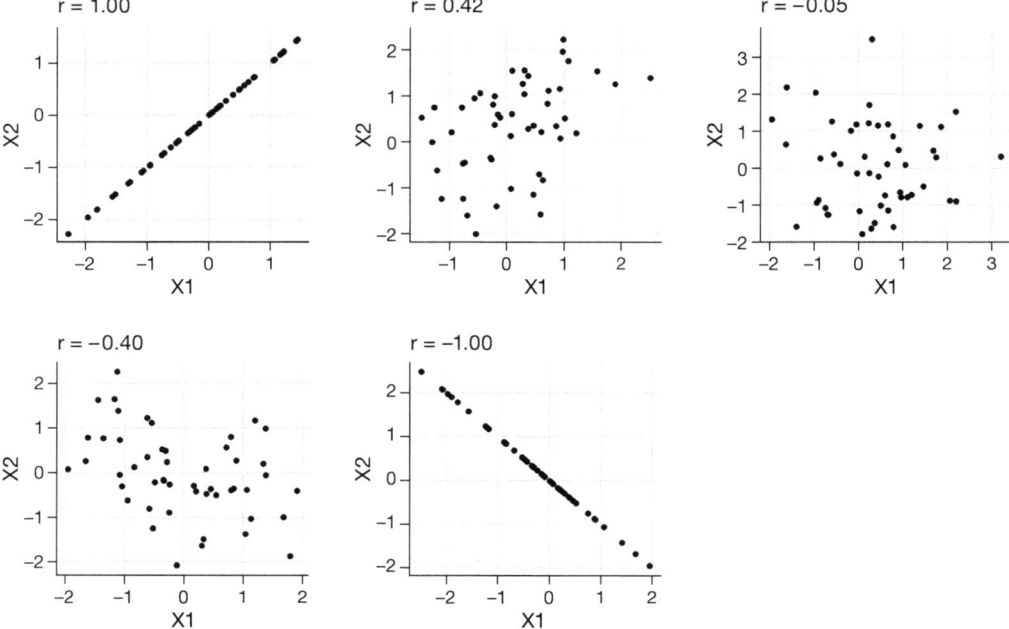

FIGURE 10.4. Examples of various levels of the correlation coefficient.

Correlation Coefficient

The correlation coefficient (r) is a measure of the strength of the linear relationship between two continuous variables. We discuss correlation in much more detail in chapter 13; here we simply introduce r as a way to quantify the relation between two variables.

r is a measure that varies from −1 to 1, where a value of 1 represents a perfect positive relationship between the variables, 0 represents no relationship, and −1 represents a perfect negative relationship. Figure 10.4 shows examples of various levels of correlation using randomly generated data.

Odds Ratio

In our earlier discussion of probability, we talked about the concept of odds—that is, the relative likelihood of some event happening versus not happening:

$$odds \ of \ A = \frac{P(A)}{P(\neg A)}$$

We also discussed the *odds ratio*, which is simply the ratio of two odds. The odds ratio is a useful way to describe effect sizes for binary variables.

For example, let's take the case of smoking and lung cancer. A study published in the *International Journal of Cancer* (Pesch et al. 2012) combined data regarding the occurrence of lung cancer in smokers and individuals who have never smoked across a number of

Table 10.2. Lung cancer occurrence separately for current smokers and those who have never smoked

Status	Never smoked	Current smoker
No cancer	2883	3829
Cancer	220	6784

different studies. Note that these data come from case-control studies, which means that participants in the studies were recruited because they either did or did not have cancer; their smoking status was then examined. These numbers (shown in table 10.2) thus do not represent the prevalence of cancer among smokers in the general population—but they can tell us about the relationship between cancer and smoking.

We can convert these numbers to odds ratios for each of the groups. The odds of a nonsmoker having lung cancer are 0.08, whereas the odds of a current smoker having lung cancer are 1.77. The ratio of these odds tells us about the relative likelihood of cancer between the two groups: the odds ratio of 23.22 tells us that the odds of lung cancer in smokers are roughly 23 times higher than never-smokers.

Statistical Power

Remember from the previous chapter that, under the Neyman-Pearson hypothesis testing approach, we have to specify our level of tolerance for two kinds of errors: false positives (which they call *Type I error*) and false negatives (which they call *Type II error*). People often focus heavily on Type I error, because making a false positive claim is generally viewed as a very bad thing; for example, the now discredited claims by Wakefield (1999) that autism was associated with vaccination led to antivaccine sentiment that has resulted in substantial increases in childhood diseases, such as measles. Similarly, we don't want to claim that a drug cures a disease if it really doesn't. That's why the tolerance for Type I errors is generally set fairly low, usually at $\alpha = 0.05$. But what about Type II errors?

The concept of *statistical power* is the complement of Type II error—that is, it is the likelihood of finding a positive result given that it exists:

$$power = 1 - \beta$$

Another important aspect of the Neyman-Pearson model that we didn't discuss earlier is the fact that, in addition to specifying the acceptable levels of Type I and Type II errors, we also have to describe a specific alternative hypothesis—that is, what is the size of the effect that we wish to detect? Otherwise, we can't interpret β—the likelihood of finding a large effect is always going to be higher than finding a small effect, so β will differ depending on the size of effect we are trying to detect.

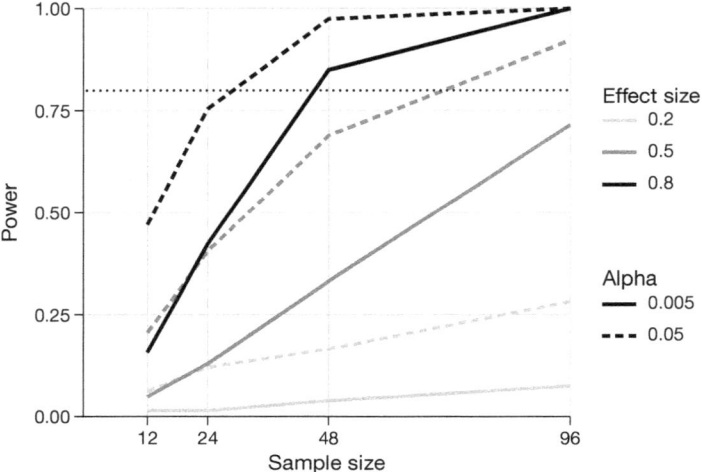

FIGURE 10.5. Results from a power simulation, showing power as a function of sample size, with effect sizes shown as different shades, and alpha shown by line type. The standard criterion of 80% power is shown by the dotted black line.

There are three factors that can affect statistical power:

- *Sample size*: Larger samples provide greater statistical power.
- *Effect size*: A given design will always have greater power to find a large effect than a small effect (because finding large effects is easier).
- *Type I error rate*: There is a relationship between Type I error and power such that (all else being equal) decreasing Type I error also decreases power.

We can see this through simulation. First, let's simulate a single experiment in which we compare the means of two groups using a standard *t* test. We vary the size of the effect (specified in terms of Cohen's d), the Type I error rate, and the sample size (per group), and for each of these we examine how the proportion of significant results (i.e., power) is affected. Figure 10.5 shows an example of how power changes as a function of these factors.

This simulation shows us that, even with a sample size of 96 in each group, we have relatively little power to find a small effect ($d = 0.2$) with $\alpha = 0.005$. This means that a study designed to do this would be futile—it is almost guaranteed to find nothing even if a true effect of that size exists.

There are at least two important reasons to care about statistical power. First, if you are a researcher, you probably don't want to spend your time doing futile experiments. Running an underpowered study essentially means that there is a very low likelihood that one will find an effect, even if it exists. Second, it turns out that any positive findings that come from an underpowered study are more likely to be false compared to a well-powered study, a point we discuss in more detail in chapter 18.

Power Analysis

Fortunately, there are tools available that allow us to determine the statistical power of an experiment. The most common use of these tools is in planning an experiment, when we would like to determine how large our sample needs to be in order to have sufficient power to find our effect of interest.

Let's say that we are interested in running a study of how a particular personality trait differs between users of iOS versus Android devices. Our plan is to collect two groups of individuals and measure them on the personality trait, and then compare the two groups using a t test. In this case, we would think that a medium effect $(d = 0.5)$ is of scientific interest, so we use that level for our power analysis. In order to determine the necessary sample size, we can use a power function from our statistical software:

```
##
##          Two-sample t-test power calculation
##
##                      n = 64
##                  delta = 0.5
##                     sd = 1
##              sig.level = 0.05
##                  power = 0.8
##            alternative = two.sided
##
## NOTE: n is number in *each* group
```

This tells us that we would need at least 64 subjects in each group in order to have sufficient power to find a medium-sized effect. It's always important to run a power analysis before one starts a new study to ensure that the study won't be futile due to a sample that is too small.

It might have occurred to you that, if the effect size is large enough, then the necessary sample will be very small. For example, if we run the same power analysis with an effect size of $d = 2$, then we will see that we only need about five subjects in each group to have sufficient power to find the difference.

```
##
##          Two-sample t-test power calculation
##
##                      n = 5.1
##                      d = 2
##              sig.level = 0.05
##                  power = 0.8
##            alternative = two.sided
##
## NOTE: n is number in *each* group
```

However, it's rare in science to be doing an experiment where we expect to find such a large effect—just as we don't need statistics to tell us that 16-year-olds are taller than than 6-year-olds. When we run a power analysis, we need to specify an effect size that is plausible and/or scientifically interesting for our study, which would usually come from previous research. However, in chapter 18 we discuss a phenomenon known as the *winner's curse* that likely results in published effect sizes being larger than the true effect size, so this should also be kept in mind.

Suggested Reading

- "Robust Misinterpretation of Confidence Intervals," by Rink Hoekstra et al. This academic publication, which appeared in the journal *Psychonomic Bulletin & Review* in 2014, outlines the many ways in which confidence intervals are misinterpreted, even by professional scientists.

Problems

1. Which of the following is true regarding confidence intervals? Choose all that apply.
   - If there is no overlap between the 95% confidence intervals for two means, then there is definitely a difference between the means at p < .05.
   - The confidence interval gets larger as the sample size gets larger.
   - The 95% confidence interval tells us the range in which there is a 95% chance that the true population parameter will fall.
   - The 95% confidence interval tells us the range that will capture the true population parameter 95% of the time.
2. A researcher computes a confidence interval that ranges from 90 to 110. Which of the following is the most appropriate interpretation of this interval?
   - There is a 95% chance that the true population mean falls between 90 and 110.
   - If 100 samples were taken from the same population, the confidence interval would capture the population mean 95% of the time.
   - We can have 95% confidence that, if we take another sample, the mean will fall between 90 and 110.
3. Select the appropriate answers from this list for the following two statements: standard deviation, standard error of the mean, variance.
   - Cohen's d involves dividing by the _____.
   - The *t* statistic involves dividing by the _____.
4. Which of the following best describes the concept of statistical power?
   - The likelihood of detecting a true effect when it exists
   - The likelihood of getting the right answer
   - The likelihood of failing to reject the null hypothesis when it is false
5. Choose the appropriate answers to complete the statements below from these options: increases, decreases, stays the same.

- Statistical power _____ as the sample size increases.
- Statistical power _____ as the effect size increases.
- Statistical power _____ as the Type I error rate increases.

6. The width of a confidence interval will (increase, decrease, stay the same) as the sample size increases.

7. Match each type of data below to the relevant effect size measure from this list: Cohen's d, correlation coefficient, odds ratio.
   - A difference between two means
   - A linear relationship between two continuous measures
   - A difference between two binary variables

8. What are three factors that can impact statistical power, and how does each factor affect power?

9. A researcher performs a power analysis, based on an effect size of $d = 0.5$. He subsequently decides that the actual effect size of interest is $d = 0.2$. Will the required sample size to find this effect be smaller or larger than the initial power analysis?

11

Bayesian Statistics

In this chapter, we take up the approach to statistical modeling and inference that stands in contrast to the null hypothesis statistical testing framework that you encountered in chapter 9. This alternative methodology is known as *Bayesian statistics*, after the Reverend Thomas Bayes, whose theorem we discussed in chapter 6. In this chapter, you will learn how Bayes' theorem provides a way of understanding data that solves many of the conceptual problems regarding null hypothesis statistical testing, while also introducing some new challenges.

Learning Objectives

Having read this chapter, you should be able to

- Describe the main differences between Bayesian analysis and null hypothesis testing.
- Describe and perform the steps in a Bayesian analysis.
- Describe the effects of different priors, and the considerations that go into choosing a prior.
- Describe the difference in interpretation between a confidence interval and a Bayesian credible interval.

Generative Models

Say you are walking down the street and a friend of yours walks right by but doesn't say hello. You would probably try to decide why this happened: Did they not see you? Are they mad at you? Are you suddenly cloaked in a magic invisibility shield? One of the basic ideas behind Bayesian statistics is that we want to infer the details of how the data are being generated based on the data themselves. In this case, you want to use the data (i.e., the fact that your friend did not say hello) to infer the process that generated the data (e.g., whether or not they actually saw you, how they feel about you).

The idea behind a generative model is that a *latent* (unseen) process generates the data we observe, usually with some amount of randomness. When we take a sample of data

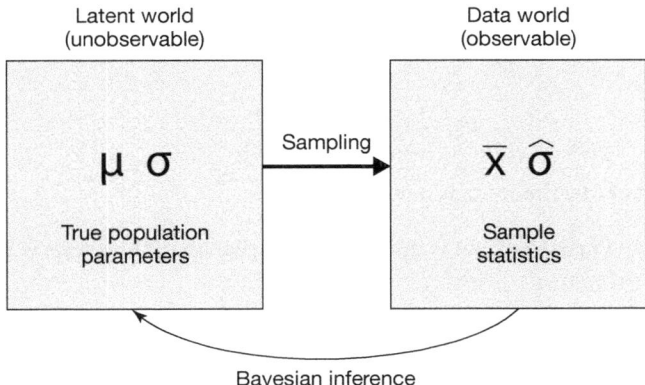

Latent world
(unobservable)

Data world
(observable)

$\mu \; \sigma$

Sampling

$\overline{x} \; \widehat{\sigma}$

True population
parameters

Sample
statistics

Bayesian inference

FIGURE 11.1. A schematic of the idea of a generative model.

from a population and estimate a parameter from the sample, what we are doing in essence is trying to learn the value of a latent variable (the population mean) that gives rise through sampling to the observed data (the sample mean). Figure 11.1 shows a schematic of this idea.

If we know the value of the latent variable, then it's easy to reconstruct what the observed data should look like. For example, let's say that we are flipping a coin that we know to be fair, such that we would expect it to land on heads 50% of the time. We can describe the coin by a binomial distribution with a value of $P_{heads} = 0.5$, and then we could generate random samples from such a distribution in order to see what the observed data should look like. However, in general we are in the opposite situation: we don't know the value of the latent variable of interest, but we have some data that we would like to use to estimate it.

Bayes' Theorem and Inverse Inference

The reason that Bayesian statistics has its name is because it takes advantage of Bayes' theorem to make inferences from data about the underlying process that generated the data. Let's say that we want to know whether a coin is fair. To test this, we flip the coin 10 times and come up with 7 heads. Before this test, we were pretty sure that $P_{heads} = 0.5$, but finding 7 heads out of 10 flips would certainly give us pause. We already know how to compute the conditional probability that we would flip 7 or more heads out of 10 if the coin is really fair $(P(n \geq 7 | p_{heads} = 0.5))$, using the binomial distribution.

The resulting probability is 0.055. That is a fairly small number, but this number doesn't really answer the question we are asking—it is telling us about the likelihood of 7 or more heads given some particular probability of heads, whereas what we really want to know is the true probability of heads for this particular coin. This should sound familiar, as it's exactly the situation that we were in with null hypothesis testing, which told us about the likelihood of data rather than the likelihood of hypotheses.

Remember that Bayes' theorem provides us with the tool we need to invert a conditional probability:

$$P(H|D) = \frac{P(D|H) * P(H)}{P(D)}$$

We can think of this theorem as having four parts:

- *Prior* ($P(hypothesis)$): What is our degree of belief about hypothesis H before seeing the data D?
- *Likelihood* ($P(data|hypothesis)$): How likely are the observed data D under hypothesis H?
- *Marginal likelihood* ($P(data)$): How likely are the observed data, combining over all possible hypotheses according to their priors?
- *Posterior* ($P(hypothesis|data)$): What is our updated belief about hypothesis H, given the data D?

In the case of our coin-flipping example:

- *Prior* (P_{heads}): What is our degree of belief about the likelihood of flipping heads, which was $P_{heads} = 0.5$?
- *Likelihood* ($P(7 \text{ or more heads out of 10 flips}|P_{heads} = 0.5)$): How likely are 7 or more heads out of 10 flips if $P_{heads} = 0.5$?
- *Marginal likelihood* ($P(7 \text{ or more heads out of 10 flips})$): How likely are we to observe 7 heads out of 10 coin flips, in general?
- *Posterior* ($P_{heads}|7 \text{ or more heads out of 10 coin flips}$): What is our updated belief about P_{heads} given the observed coin flips?

Here we see one of the primary differences between frequentist and Bayesian statistics. Frequentists do not believe in the idea of a probability of a hypothesis (i.e., our degree of belief about a hypothesis)—for them, either a hypothesis is true or it isn't. Another way to say this is that, for the frequentist, the hypothesis is fixed and the data are random, which is why frequentist inference focuses on describing the probability of data given a hypothesis (i.e., the p-value). Bayesians, on the other hand, are comfortable making probability statements about both data and hypotheses.

Doing Bayesian Estimation

We ultimately want to use Bayesian statistics to make decisions about hypotheses, but before we do that, we need to estimate the parameters that are necessary to make the decision. Here we will walk through the process of Bayesian estimation. Let's use another screening example: airport security screening. If you fly a lot, it's just a matter of time until one of the random explosive screenings comes back positive; I had the particularly unfortunate experience of this happening soon after September 11, 2001, when airport security staff were especially on edge.

What the security staff want to know is the likelihood that a person is carrying an explosive, given that the machine has given a positive test. Let's walk through how to calculate this value using Bayesian analysis.

Specifying the Prior

To use Bayes' theorem, we first need to specify the prior probability for the hypothesis. In this case, we don't know the real number, but we can assume that it's quite small. According to the FAA, there were 971,595,898 air passengers in the US in 2017. Let's say that one of those travelers was carrying an explosive in their bag—that would give a prior probability of 1 out of 971 million, which is very small! The security personnel may have reasonably held a stronger prior in the months after the 9/11 attack, so let's say that their subjective belief was that one out of every million flyers was carrying an explosive.

Collect Some Data

The data are composed of the results of the explosive screening test. Let's say that the security staff runs the bag through their testing apparatus three times, and it gives a positive reading on three of the three tests.

Computing the Likelihood

We want to compute the likelihood of the data under the hypothesis that there is an explosive in the bag. Let's say that we know (from the machine's manufacturer) that the *sensitivity* of the test is 0.99—that is, when a device is present, it will detect it 99% of the time. To determine the likelihood of our data under the hypothesis that a device is present, we can treat each test as a Bernoulli trial (ie., a trial with an outcome of true or false) with a probability of success of 0.99, which we can model using a binomial distribution.

Computing the Marginal Likelihood

We also need to know the overall likelihood of the data—that is, finding three positives out of three tests—weighting each possible outcome by its prior. Computing the marginal likelihood is often one of the most difficult aspects of Bayesian analysis, but for our example it's simple because we can take advantage of the specific form of Bayes' theorem for a binary outcome that we introduced in chapter 6:

$$P(E|T) = \frac{P(T|E) * P(E)}{P(T|E) * P(E) + P(T|\neg E) * P(\neg E)}$$

where E refers to the presence of explosives and T refers to a positive test result.

The marginal likelihood in this case is a weighted average of the likelihood of the data under either presence or absence of the explosive, multiplied by the probability of the explosive being present (i.e., the prior) or absent ($1 -$ the prior), respectively. In this case, let's say that we know (from the manufacturer) that the specificity of the test is 0.99, such that the likelihood of a positive result when there is no explosive ($P(T|\neg E)$) is 0.01.

Computing the Posterior

We now have all the parts that we need to compute the posterior probability of an explosive being present, given the observed three positive outcomes out of three tests. This result shows us that the posterior probability of an explosive in the bag given these positive tests (0.492) is just under 50%, again highlighting the fact that testing for rare events is almost always liable to produce high numbers of false positives, even when the specificity and sensitivity are very high.

  An important aspect of Bayesian analysis is that it can be sequential. Once we have the posterior from one analysis, it can become the prior for the next analysis!

Estimating Posterior Distributions

In the previous example, there were only two possible outcomes—either the explosive is there or it's not—and we wanted to know which outcome was most likely, given the data. However, in other cases we want to use Bayesian analysis to estimate the numeric value of a parameter. Let's say that we want to know about the effectiveness of a new drug for pain; to test this, we can administer the drug to a group of patients and then ask them whether their pain was improved or not after taking the drug. We can use Bayesian analysis to estimate the proportion of people for whom the drug will be effective using these data.

Specifying the Prior

In this case, we don't have any prior information about the effectiveness of the drug, so we use a *uniform distribution* as our prior, since all values are equally likely under a uniform distribution. In order to simplify the example, we only look at a subset of 99 possible values of effectiveness (from .01 to .99, in steps of .01). Therefore, each possible value has a prior probability of 1/99. The use of discrete possible values greatly simplifies the Bayesian computation, letting us avoid the use of calculus.

Collect Some Data

We need some data in order to estimate the effect of the drug. Let's say that we administer the drug to 100 individuals and we find that 64 respond positively to the drug.

Computing the Likelihood

We can compute the likelihood of the observed data under any particular value of the effectiveness parameter using the binomial density function. In figure 11.2, you can see the likelihood curves over numbers of responders for several different values of $P_{respond}$. Looking at this figure, it seems that our observed data are relatively more likely under the

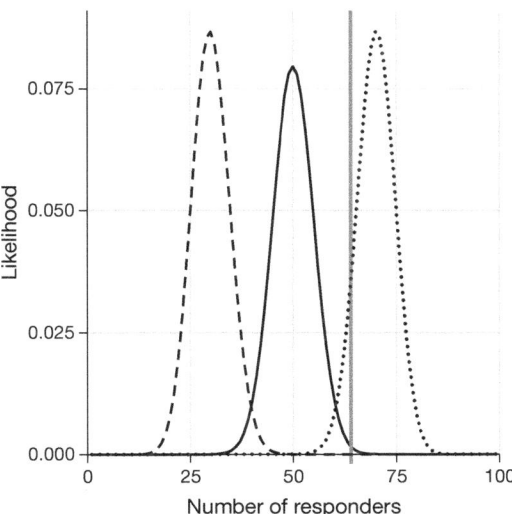

FIGURE 11.2. Likelihood of each possible number of responders under several different hypotheses $p_{respond} = 0.5$ (solid), 0.7 (dotted), 0.3 (dashed). The observed value is indicated by the vertical line.

hypothesis of $P_{respond} = 0.7$, somewhat less likely under the hypothesis of $P_{respond} = 0.5$, and quite unlikely under the hypothesis of $P_{respond} = 0.3$. One of the fundamental ideas of Bayesian inference is that we should upweight our belief in values of our parameter of interest in proportion to how likely the data are under those values, balanced against what we believed about the parameter values before having seen the data (our prior knowledge).

Computing the Marginal Likelihood

In addition to the likelihood of the data under different hypotheses, we need to know the overall likelihood of the data, combining across all hypotheses (i.e., the marginal likelihood). This marginal likelihood is primarily important because it helps ensure that the posterior values are true probabilities. In this case, our use of a set of discrete possible parameter values makes it easy to compute the marginal likelihood, because we can just compute the likelihood of each parameter value under each hypothesis and add them up.

Computing the Posterior

We now have all the parts that we need to compute the posterior probability distribution across all possible values of $p_{respond}$, as shown in figure 11.3.

Maximum a Posteriori (MAP) Estimation

Given our data, we would like to obtain an estimate of $p_{respond}$ for our sample. One way to do this is to find the value of $p_{respond}$ for which the posterior probability is the highest, which we refer to as the *maximum a posteriori* (MAP) estimate. We can find this value from

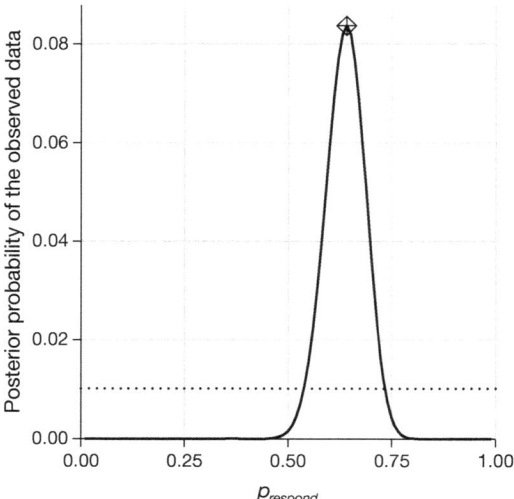

FIGURE 11.3. Posterior probability distribution for the observed data (solid line) against the uniform prior distribution (dotted line). The maximum a posteriori (MAP) value is signified by the diamond.

the data in figure 11.3—it's the value shown with a marker at the top of the distribution. Note that the result (0.64) is simply the proportion of responders from our sample—this occurs because the prior was uniform and thus didn't influence our estimate.

Credible Intervals

Often we would like to know not just a single estimate for the posterior, but an interval in which we are confident that the posterior falls. We previously discussed the concept of confidence intervals in the context of frequentist inference, and you may remember that the interpretation of confidence intervals was particularly convoluted: it is an interval that contains the true value of the parameter 95% of the time. What we really want is an interval in which we are confident that the true parameter falls, and Bayesian statistics can give us such an interval, which we call a *credible interval*.

The interpretation of this credible interval is much closer to what we had hoped we could get from a confidence interval (but could not): it tells us that there is a 95% probability that the value of $p_{respond}$ falls between these two values. Importantly, in this case it shows that we have high confidence that $p_{respond} > 0.0$, meaning that the drug seems to have a positive effect.

In some cases, the credible interval can be computed *numerically* based on a known distribution, but it's more common to generate a credible interval by sampling from the posterior distribution and then to compute quantiles of the samples. This is particularly useful when we don't have an easy way to express the posterior distribution numerically, which is often the case in real Bayesian data analysis. One such method (rejection sampling) is explained in more detail in the appendix at the end of this chapter.

Effects of Different Priors

In the previous example, we used a *flat prior*, meaning that we didn't have any reason to believe that any particular value of $p_{respond}$ was more or less likely. However, let's say that we had instead started with some previous data: In a previous study, researchers had tested 20 people and found that 10 of them had responded positively. This would led us to start with a prior belief that the treatment has an effect in 50% of people. We can do the same computation as above, but using the information from our previous study to inform our prior (panel A in figure 11.4). Note that the likelihood and marginal likelihood did not change—only the prior changed. The effect of the change in the prior to was to pull the posterior closer to the mass of the new prior, which is centered at 0.5.

Now let's see what happens if we come to the analysis with an even stronger prior belief. Let's say that, instead of having previously observed 10 responders out of 20 people, the prior study had tested 500 people and found 250 responders. This should in principle give us a much stronger prior, and as we see in panel B of figure 11.4 , that's what happens: the prior is much more concentrated around 0.5, and the posterior is also much closer to the prior. The general idea is that Bayesian inference combines the information from the prior and the likelihood, weighting the relative strength of each. This example also highlights the sequential nature of Bayesian analysis—the posterior from one analysis can become the prior for the next analysis.

Finally, it is important to realize that, if the priors are strong enough, they can completely overwhelm the data. Let's say that you have an absolute prior that $p_{respond}$ is 0.8 or greater, such that you set the prior likelihood of all other values to zero. What happens if we then compute the posterior? In panel C of figure 11.4, we see that there is zero density in the posterior for any of the values where the prior was set to zero—the data are overwhelmed by the absolute prior.

Choosing a Prior

The impact of priors on the resulting inferences are the most controversial aspect of Bayesian statistics. What is the right prior to use? If the choice of the prior determines the results (i.e., the posterior), how can you be sure your results are trustworthy? These are difficult questions, but we should not back away just because we are faced with hard questions. As we discussed previously, Bayesian analyses give us interpretable results (e.g., credible intervals). This alone should inspire us to think hard about these questions so that we can arrive at results that are reasonable and interpretable.

There are various ways to choose one's priors, which (as we saw above) can impact the resulting inferences. Sometimes we have a very specific prior, as in the case where we expected our coin to land heads 50% of the time, but in many cases we don't have such a strong starting point. *Uninformative priors* attempt to influence the resulting posterior as little as possible, as we saw in the example of the uniform prior above. It's also common to use *weakly informative priors* (or *default priors*), which influence the result only very slightly. For example, if we had used a binomial distribution based on one heads out of

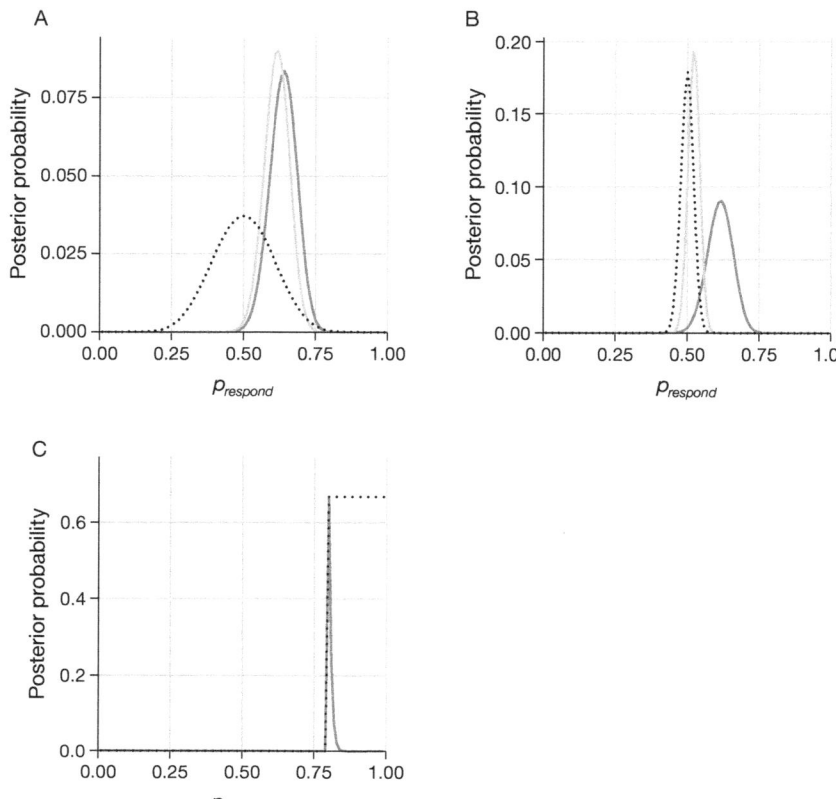

FIGURE 11.4. Effects of priors on the posterior distribution. (A) The original posterior distribution based on a flat prior is plotted in dark gray. The prior based on the observation of 10 responders out of 20 people is plotted in the dotted line, and the posterior using this prior is plotted in light gray. (B) Effects of the strength of the prior on the posterior distribution. The dark gray line shows the posterior obtained using the prior based on 50 responders out of 100 people. The dotted line shows the prior based on 250 responders out of 500 people, and the light gray line shows the posterior based on that prior. (C) Effects of the strength of the prior on the posterior distribution. The dark gray line shows the posterior obtained using an absolute prior, which states that $p_{respond}$ is 0.8 or greater. The prior is shown by the dotted line.

two coin flips, the prior would have been centered around 0.5 but fairly flat, influencing the posterior only slightly. It is also possible to use priors based on the scientific literature or preexisting data, which we would call *empirical priors*. In general, however, we will stick to the use of uninformative/weakly informative priors, since they raise the least concern about influencing our results.

Bayesian Hypothesis Testing

Having learned how to perform Bayesian estimation, we now turn to the use of Bayesian methods for hypothesis testing. Let's say that there are two politicians who differ in their beliefs about whether the public is in favor of an extra tax to support the national parks. Senator Smith thinks that only 40% of people are in favor of the tax, whereas Senator Jones thinks that 60% of people are in favor. They arrange to have a poll done to test this, which asks 1000 randomly selected people whether they support such a tax. The results are that 490 of the people in the polled sample were in favor of the tax. Based on these data, we would like to know the following: Do the data support the claims of one senator over the other, and by how much? We can test this using a concept known as the *Bayes factor*, which quantifies which hypothesis is better by comparing how well each predicts the observed data.

Bayes Factors

The Bayes factor characterizes the relative likelihood of the data under two different hypotheses. It is defined as

$$BF = \frac{p(data|H_1)}{p(data|H_2)}$$

for two hypotheses H_1 and H_2, assuming that their prior probabilities are equal. In the case of our two senators, we know how to compute the likelihood of the data under each hypothesis using the binomial distribution; let's assume for the moment that our prior probability for each senator being correct is the same ($P_{H_1} = P_{H_2} = 0.5$). We put Senator Smith in the numerator and Senator Jones in the denominator, so that a value greater than one will reflect greater evidence for Senator Smith, and a value less than one will reflect greater evidence for Senator Jones. The resulting Bayes factor (3325.26) provides a measure of the evidence that the data provide regarding the two hypotheses—in this case, it tells us that Senator Smith's hypothesis is much more strongly supported by the data than Senator Jones's hypothesis.

Bayes Factors for Statistical Hypotheses

In the previous example, we had specific predictions from each senator, whose likelihood we could quantify using the binomial distribution. In addition, our prior probability for the two hypotheses was equal. However, in real data analysis we generally must deal with uncertainty about our parameters, which complicates the Bayes factor, because we need to compute the marginal likelihood—that is, an integrated average of the likelihoods over all possible model parameters, weighted by their prior probabilities. However, in exchange we gain the ability to quantify the relative amount of evidence in favor of the null versus alternative hypotheses.

Let's say that we are a medical research team performing a clinical trial for the treatment of diabetes, and we wish to know whether a particular drug reduces blood glucose compared to a placebo. We recruit a set of volunteers and randomly assign them to either drug or placebo group, and we measure the change in hemoglobin A1C (a marker for blood glucose levels) in each group over the period in which the drug or placebo was administered. What we want to know is, Is there a difference between the drug and the placebo?

First, let's generate some data and analyze them using null hypothesis testing (figure 11.5). Then let's perform an independent-samples *t* test, which shows that there is a significant difference between the groups:

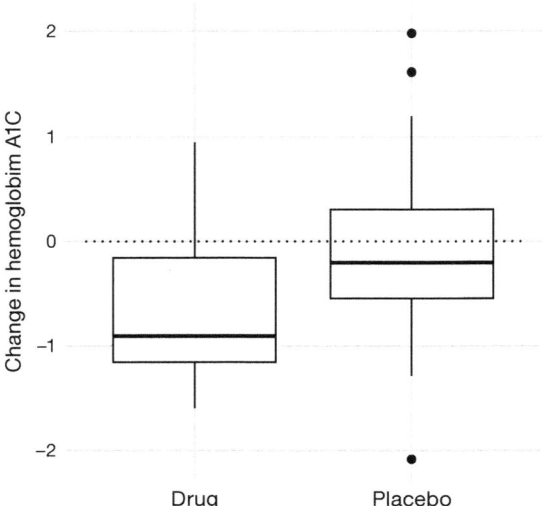

FIGURE 11.5. Box plots showing data for drug and placebo groups.

```
##
##   Welch Two-Sample t-test
##
## data:   hbchange by group
## t = 2, df = 32, p-value = 0.02
## alternative hypothesis: true difference in means
## between group 0 and group 1 is greater than 0

## sample estimates:
## mean in group 0    mean in group 1
## -0.082             -0.650
```

This test tells us that there is a significant difference between the groups, but it doesn't quantify how strongly the evidence supports the null versus alternative hypotheses. To measure that, we can compute a Bayes factor using the `ttestBF` function from the BayesFactor package in R:

```
## Bayes factor analysis
## --------------
## [1] Alt., r=0.707 0<d<Inf     : 3.4   ±0%
## [2] Alt., r=0.707 !(0<d<Inf) : 0.12 ±0.01%
##
## Against denominator:
##   Null, mu1-mu2 = 0
```

```
## ---
## Bayes factor type: BFindepSample, JZS
```

We are particularly interested in the Bayes factor for an effect greater than zero, which is listed in the line marked [1] in the report. The Bayes factor here (3.4) tells us that the alternative hypothesis (i.e., that the difference is greater than or less than zero) is about three times more likely than the point null hypothesis (i.e., a mean difference of exactly zero) given the data. Thus, while the effect is significant, the amount of evidence it provides us in favor of the alternative hypothesis is rather weak.

ONE-SIDED TESTS

We generally are less interested in testing against the null hypothesis of a specific point value (e.g., mean difference = 0) than we are in testing against a directional null hypothesis (e.g., that the difference is less than or equal to zero). We can also perform a directional (or *one-sided*) test using the results from the ttestBF analysis, since it provides two Bayes factors: one for the alternative hypothesis that the mean difference is greater than zero, and one for the alternative hypothesis that the mean difference is less than zero. If we want to assess the relative evidence for a positive effect, we can compute a Bayes factor comparing the relative evidence for a positive versus a negative effect by simply dividing the two Bayes factors returned by the function:

```
## Bayes factor analysis
## --------------
## [1] Alt., r=0.707 0<d<Inf : 29 ±0.01%
##
## Against denominator:
##   Alternative, r = 0.707106781186548, mu =/= 0 !(0<d<Inf)
## ---
## Bayes factor type: BFindepSample, JZS
```

Now we see that the Bayes factor for a positive effect versus a negative effect is substantially larger (almost 30).

INTERPRETING BAYES FACTORS

How do we know whether a Bayes factor of 2 or 20 is good or bad? There is a general guideline for interpretation of Bayes factors suggested by Kass and Raftery (1995):

Bayes factor	Strength of evidence
1 to 3	Not worth more than a bare mention
3 to 20	Positive
20 to 150	Strong
>150	Very strong

Based on this table, even though the statistical result is significant, the amount of evidence in favor of the alternative versus the point null hypothesis is weak enough that it's hardly worth even mentioning, whereas the evidence for the directional hypothesis is relatively strong.

Assessing Evidence for the Null Hypothesis

Because the Bayes factor compares evidence for two hypotheses, it also allows us to assess whether there is evidence in favor of the null hypothesis, which we couldn't do with standard null hypothesis testing (because it starts with the assumption that the null is true). This can be very useful for determining whether a nonsignificant result really provides strong evidence that there is no effect or instead just reflects weak evidence overall.

Suggested Reading

- *The Theory That Would Not Die: How Bayes' Rule Cracked the Enigma Code, Hunted Down Russian Submarines, and Emerged Triumphant from Two Centuries of Controversy,* by Sharon Bertsch McGrayne. An outstanding introduction to the ideas behind Bayesian statsitics.
- *Doing Bayesian Data Analysis: A Tutorial Introduction with R,* by John K. Kruschke. A solid guide to the practice of Bayesian data analysis, with examples using the R programming language.

Problems

1. Choose the proper responses for the following statements:
   - Frequentists (do/do not) believe in the idea of a probability of a hypothesis.
   - Bayesians (do/do not) believe in the idea of a probability of a hypothesis.
2. Match each statement below to one of these two options: a hypothesis given the observed data, the observed data under the null hypothesis.
   - Frequentist statistics focus on estimating the probability of _____.
   - Bayesian statistics focus on estimating the probability of _____.
3. A research study compares a new treatment for Alzheimer's disease versus a placebo and reports a Bayes factor of 2.8 for the effect of the treatment.
   - First, explain in your own words what a Bayes factor is.
   - Second, explain how you would interpret the specific value of the Bayes factor reported for this study.
4. What can happen if the prior is strong enough?
   - It can overwhelm the data.
   - It causes the result to be identical to a frequentist analysis.
   - It causes Bayesian estimation to fail.
5. Which of the following are important differences between Bayesian and frequentist analyses? Choose all that apply.
   - Bayesian analysis allows one to quantify evidence in favor of or against the null hypothesis.

- Bayesian analysis can be biased by subjective priors.
- Bayesian analysis estimates the likelihood of the data under the particular hypothesis.
- Bayesian analysis allows one to determine whether a nonsignificant result reflects a truly negative result versus insufficient evidence.
6. How can Bayesian inference be considered as a form of learning?

Appendix

Rejection Sampling

We can generate samples from our posterior distribution using a simple algorithm known as *rejection sampling*. The idea is that we choose a random value of x (in this case, $p_{respond}$) and a random value of y (in this case, the posterior probability of $p_{respond}$) each from a uniform distribution. We then only accept the sample if $y < f(x)$—in this case, if the randomly selected value of y is less than the actual posterior probability of y. Figure 11.6 shows an example of a histogram of samples using rejection sampling, along with the 95% credible interval obtained using this method (with the values presented in table 11.1).

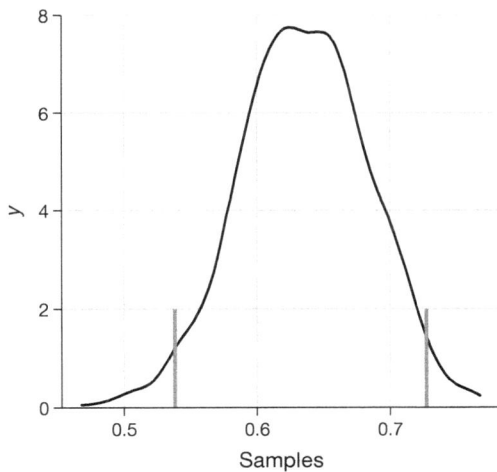

FIGURE 11.6. Rejection sampling example. The curve shows the density of all possible values of $p_{respond}$; the vertical lines show the 2.5 and 97.5 percentiles of the distribution, which represent the 95% credible interval for the estimate of $p_{respond}$.

Table 11.1. Credible Interval Obtained for the Pain Drug Example using Rejection Sampling

Percentile	Value of x
2.5%	0.54
97.5%	0.73

12

Modeling Categorical Relationships

So far we have discussed the general concepts of statistical modeling and hypothesis testing, and applied them to some simple analyses; in the next few chapters, we turn to the question of how to model and test for particular kinds of relationships in our data. In this chapter, we focus on the modeling of *categorical* relationships, by which we mean relationships between variables that are measured qualitatively. These data are usually expressed in terms of counts; that is, for each value of the variable (or combination of values of multiple variables), how many observations take that value? For example, when we count how many people from each major are in our class, we are fitting a categorical model to the data.

Learning Objectives

Having read this chapter, you should be able to

- Describe the concept of a contingency table for categorical data.
- Describe the concept of the chi-squared test for association and compute it for a given contingency table.
- Describe Simpson's paradox and why it is important for categorical data analysis.

Candy Colors: An Example

Let's say that I have purchased a bag of 100 candies, which are labeled as having 1/3 chocolates, 1/3 licorices, and 1/3 gumballs. When I count the candies in the bag, we get the following numbers: 30 chocolates, 33 licorices, and 37 gumballs. Because I like chocolate much more than licorice or gumballs, I feel slightly ripped off; however, I know that there is a bit of error in the candy sorting machines, and I'd like to know if this was just a random accident. To answer that question, I need to know the following: What is the likelihood that the count would come out this way if the true probability of each candy type is the averaged proportion of 1/3 each?

Table 12.1. Observed Counts, Expectations under the Null Hypothesis, and Squared Differences in the Candy Data

Candy type	Count	Null expectation	Squared difference
Chocolate	30	33	11.11
Licorice	33	33	0.11
Gumball	37	33	13.44

The Chi-Squared Test

The chi-squared test provides us with a way to test whether a set of observed counts differs from some specific expected values that define the null hypothesis:

$$\chi^2 = \sum_i \frac{(observed_i - expected_i)^2}{expected_i}$$

In the case of our candy example, the null hypothesis is that the proportion of each type of candy is equal. To compute the chi-squared statistic, we first need to come up with our expected counts under the null hypothesis: since the null is that they are all the same, then this is just the total count split across the three categories (as shown in table 12.1). We then take the difference between each count and its expectation under the null hypothesis, square them, divide them by the null expectation, and add them up to obtain the chi-squared statistic.

The chi-squared statistic for this analysis comes out to 0.74, which on its own is not interpretable, since it depends on the number of different values that were added together. However, we can take advantage of the fact that the chi-squared statistic is distributed according to a specific distribution under the null hypothesis, which is known as the *chi-squared distribution*. This distribution is defined as the sum of squares of a set of standard normal random variables; it has a number of degrees of freedom that is equal to the number of variables being added together. The shape of the distribution depends on the number of degrees of freedom. Panel A of figure 12.1 shows examples of the distribution for several different degrees of freedom.

Let's verify that the chi-squared distribution accurately describes the sum of squares of a set of standard normal random variables, using simulation. To do this, we repeatedly draw sets of eight random numbers and add up each set after squaring each value. Panel B of figure 12.1 shows that the theoretical distribution matches closely with the results of a simulation that repeatedly added together the squares of a set of random normal variables.

For the candy example, we can compute the likelihood of our observed chi-squared value of 0.74 or greater under the null hypothesis of equal frequency across all candies. We use a chi-squared distribution with degrees of freedom equal to $k - 1$ (where $k =$ the number of categories) since we lost one degree of freedom when we computed the mean

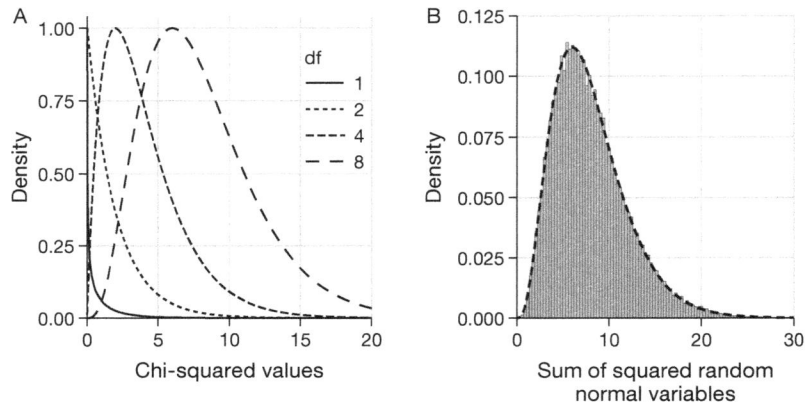

FIGURE 12.1. (A) Examples of the chi-squared distribution for various degrees of freedom. (B) Simulation of sum of squared random normal variables. The histogram is based on the sum of squares of 50,000 sets of eight random normal variables; the dotted line shows the values of the theoretical chi-squared distribution with eight degrees of freedom.

in order to generate the expected values. The resulting p-value $(P(\chi^2 > 0.74) = 0.691)$ shows that the observed counts of candies are not particularly surprising based on the proportions printed on the bag of candy, and we would retain the null hypothesis of equal proportions.

Contingency Tables and the Two-Way Test

Another way that we often use the chi-squared test is to ask whether two categorical variables are related to one another. As a more realistic example, let's take the question of whether a Black driver is more likely to be searched when pulled over by a police officer, compared to a White driver. The Stanford Open Policing Project (https://openpolicing.stanford.edu/) has studied this question and provides data that we can use to analyze it. We use the data from the state of Connecticut since it is fairly small and thus easier to analyze.

The standard way to represent data from a categorical analysis is through a *contingency table*, which presents the number or proportion of observations falling into each possible combination of values for each of the variables. Table 12.2 shows the contingency table for the police search data. It can also be useful to look at the contingency table using proportions rather than raw numbers since they are easier to compare visually, so we include both absolute and relative numbers here.

The chi-squared test allows us to test whether observed frequencies are different from expected frequencies, so we need to determine what frequencies we would expect in each cell if searches and race were unrelated—by which we mean that they are *statistically independent*. Remember from chapter 6 that if X and Y are independent, then

$$P(X \cap Y) = P(X) * P(Y)$$

Table 12.2. Contingency Table for Police Search Data

Searched	Black	White	Black (relative)	White (relative)
False	36,244	239,241	0.130	0.855
True	1219	3108	0.004	0.011

That is, the joint probability under the null hypothesis of independence is simply the product of the *marginal* probabilities of each individual variable. The marginal probabilities are the probabilities of each event occurring regardless of other events. We can compute those marginal probabilities and then multiply them together to get the expected proportions under independence.

	Black	White	
Not searched	P(NS) * P(B)	P(NS) * P(W)	P(NS)
Searched	P(S) * P(B)	P(S) * P(W)	P(S)
	P(B)	P(W)	

We then compute the chi-squared statistic, which comes out to 828.3. To compute a p-value, we need to compare it to the null chi-squared distribution in order to determine how extreme our chi-squared value is compared to our expectation under the null hypothesis. The degrees of freedom for this distribution are $df = (nRows - 1) * (nColumns - 1)$—thus, for a 2×2 table like the one here, $df = (2 - 1) * (2 - 1) = 1$. The intuition here is that computing the expected frequencies requires us to use three values: the total number of observations and the marginal probability for each of the two variables. Thus, once those values are computed, there is only one number that is free to vary, and thus there is one degree of freedom. Given this, we can compute the p-value for the chi-squared statistic, which is about as close to zero as one can get: $3.79 * 10^{-182}$. This shows that the observed data would be highly unlikely if there were truly no relationship between race and police searches, and thus we should reject the null hypothesis of independence.

We can also perform this test easily using our statistical software:

```
##
##   Pearson's Chi-squared test
##
## data:  summaryDf2wayTable and 1
## X-squared = 828, df = 1, p-value <2e-16
```

Standardized Residuals

When we find a significant effect with the chi-squared test, this tells us that the data are unlikely under the null hypothesis, but it doesn't tell us *how* the data differ. To get a deeper

Table 12.3. Summary of Standardized Residuals for Police Search Data

Searched	Driver's race	Standardized residuals
False	Black	−3.3
True	Black	26.6
False	White	1.3
True	White	−10.4

insight into how the data differ from what we would expect under the null hypothesis, we can examine the residuals from the model, which reflect the deviation of the data (i.e., the observed frequencies) from the model (i.e., the expected frequencies) in each cell. Rather than looking at the raw residuals (which depend on the number of observations in the data), it's more common to look at the *standardized residuals*, which are computed as

$$standardized\ residual_{ij} = \frac{observed_{ij} - expected_{ij}}{\sqrt{expected_{ij}}}$$

where i and j are the indices for the rows and columns, respectively.

Table 12.3 shows the residuals for the police search data. These standardized residuals can be interpreted as Z-scores—in this case, we see that the number of searches for Black individuals is substantially higher than expected based on independence, and the number of searches for White individuals is substantially lower than expected. This provides us with the context that we need to interpret the significant chi-squared result.

Odds Ratios

We can also represent the relative likelihood of different outcomes in the contingency table using the *odds ratio*, which we introduced earlier, in order to better understand the magnitude of the effect. First, we represent the odds of being stopped for each race; then we compute their ratio:

$$odds_{searched|Black} = \frac{N_{searched \cap Black}}{N_{not\ searched \cap Black}} = \frac{1219}{36,244} = 0.034$$

$$odds_{searched|White} = \frac{N_{searched \cap White}}{N_{not\ searched \cap White}} = \frac{3108}{23,9241} = 0.013$$

$$odds\ ratio = \frac{odds_{searched|Black}}{odds_{searched|White}} = 2.59$$

The odds ratio shows that the odds of being searched are 2.59 times higher for Black versus White drivers, based on this dataset.

Table 12.4. Relationship between Depression
and Sleep Problems in the NHANES Dataset

Depressed	No sleep trouble	Sleep trouble
None	2614	676
Several	418	249
Most	138	145

Bayes Factor

We discussed Bayes factors in chapter 11—you may remember that they represent the ratio of the likelihood of the data under each of the two hypotheses:

$$K = \frac{P(data|H_A)}{P(data|H_0)} = \frac{P(H_A|data) * P(H_A)}{P(H_0|data) * P(H_0)}$$

We can compute the Bayes factor for the police search data using our statistical software:

```
## Bayes factor analysis
## --------------
## [1] Non-indep. (a=1) : 1.8e+142 ±0%
##
## Against denominator:
##   Null, independence, a = 1
## ---
## Bayes factor type: BFcontingencyTable, independent multinomial
```

This shows that the evidence in favor of a relationship between the driver's race and police searches in this dataset is exceedingly strong—$1.8 * 10^{142}$ is about as close to infinity as we can imagine getting in statistics.

Categorical Analysis beyond the 2 × 2 Table

Categorical analysis can also be applied to contingency tables where there are more than two categories for each variable. For example, let's look at the NHANES dataset and the variable *Depressed*, which denotes the "self-reported number of days where the participant felt down, depressed or hopeless." The answers are recorded as "none," "several," or "most." Now let's test whether this variable is related to the *SleepTrouble* variable, which denotes whether the individual has reported sleeping problems to a doctor.

Simply by looking at these data (table 12.5), we can tell that it is likely that there is a relationship between the two variables; notably, while the total number of people with sleep trouble is much less than those without, for people who report being depresssed "most" days, the number with sleep problems is greater than those without. We can quantify this directly using the chi-squared test:

```
##
##   Pearson's Chi-squared test
##
## data:  depressedSleepTroubleTable
## X-squared = 191, df = 2, p-value <2e-16
```

This test shows that there is a strong relationship between depression and sleep trouble. We can also compute the Bayes factor to quantify the strength of the evidence in favor of the alternative hypothesis:

```
## Bayes factor analysis
## -------------
## [1] Non-indep. (a=1) : 1.8e+35 ±0%
##
## Against denominator:
##   Null, independence, a = 1
## ---
## Bayes factor type: BFcontingencyTable, joint multinomial
```

Here we see that the Bayes factor is very large $(1.8 * 10^{35})$, showing that the evidence in favor of a relation between depression and sleep problems is very strong.

Beware of Simpson's Paradox

The contingency tables presented above represent summaries of large numbers of observations, but summaries can sometimes be misleading. Let's take an example from baseball. The table below shows the batting data (hits/at-bats and batting average) for Derek Jeter and David Justice over the years 1995–1997:

Player	1995		1996		1997		Combined	
Jeter	12/48	.250	183/582	.314	190/654	.291	385/1284	**.300**
Justice	104/411	**.253**	45/140	**.321**	163/495	**.329**	312/1046	.298

If you look closely, you will see that something odd is going on: in each individual year, Justice had a higher batting average than Jeter, but when we combine the data across all three years, Jeter's average is actually higher than Justice's! This is an example of a phenomenon known as *Simpson's paradox*, in which a pattern that is present in a combined dataset may not be present in any of the subsets of the data. This occurs when there is another variable that may be changing across the different subsets—in this case, the number of at-bats varies across years, with Justice batting many more times in 1995 (when batting averages were low). We refer to this as a *lurking variable*, and it's always important to be attentive to such variables whenever one examines categorical data.

Suggested Reading

- "Simpson's Paradox in Psychological Science: A Practical Guide," by R. A. Kievit et al. This academic paper, which appeared in the journal *Frontiers in Psychology* in 2013, outlines the ways in which Simpson's paradox can occur in real data.

Problems

1. Which of the following best describes the goal of the chi-squared test?
   - Comparing means between a number of groups
   - Comparing observed and expected counts
   - Comparing a continuous variable to the normal distribution
   - Comparing two continuous variables
2. How is the chi-squared distribution defined, and how are the degrees of freedom determined?
3. A poll identifies 100 people with children and 100 without children, and asks each of them whether they support a proposed ballot initiative for higher taxes to support public schools. They find that, of the people with children, 83 support the initiative, while for the people without children, 67 support the initiative.
   - Create a contingency table based on these results.
   - What are the expected values for the contingency table if support for the initiative is independent of having children?
   - Perform a chi-squared test to assess the null hypothesis of independence.
   - How do you interpret the result of this test?
4. Why does the chi-squared test for independence in a two-way contingency table only have one degree of freedom?
5. How is the standardized residual useful for interpreting the result of a chi-squared test?
6. Using the numbers from question 3, compute the standardized residuals and explain how they inform your interpretation of the result.
7. An analysis of admissions to a university shows that overall there are more females admitted than males, but when each department is examined separately, there are more males than females admitted. What phenomenon would this be an example of?
   - Type II error
   - The law of small numbers
   - Reverse causation
   - Simpson's paradox

8. Let's say that I am a café owner and I would like to test whether the number of customers at my café varies across weekdays. I count the number of customers each day for six weeks and come up with the following counts:

   Monday: 127 Tuesday: 129 Wednesday: 125 Thursday: 136 Friday: 161

   • Compute the chi-squared statistic to test the null hypothesis that the number of customers is equal across all days.

   • Based on the result of this test, is there a significant difference between days (at $p < .05$)?

13

Modeling Continuous Relationships

Most people are familiar with the idea of *correlation*, and in this chapter we provide a more formal understanding for this commonly used and misunderstood concept.

Learning Objectives

Having read this chapter, you should be able to

- Describe the concept of the correlation coefficient and its interpretation.
- Compute the correlation between two continuous variables.
- Describe the effect of outlier data points and how to address them.
- Describe the potential causal influences that can give rise to an observed correlation.

Hate Crimes and Income Inequality: An Example

In 2017, the website fivethirtyeight.com published a story titled "Higher Rates of Hate Crimes Are Tied to Income Inequality," which discussed the relationship between the prevalence of hate crimes and income inequality in the wake of the 2016 presidential election. The story reported an analysis of hate crime data from the FBI and the Southern Poverty Law Center, on the basis of which they reported the following:

> We found that income inequality was the most significant determinant of population-adjusted hate crimes and hate incidents across the United States. (Majumder 2017)

The data for this analysis are available as part of the `fivethirtyeight` package for the R statistical software, which makes it easy for us to access them. The analysis reported in the story focused on the relationship between income inequality (defined by a quantity called the *Gini index*—see the chapter appendix for more details) and the prevalence of hate crimes in each state. The relationship between income inequality and rates of hate crimes is shown in figure 13.1. Looking at the data, it seems that there may be a positive relationship between the two variables. How can we quantify that relationship?

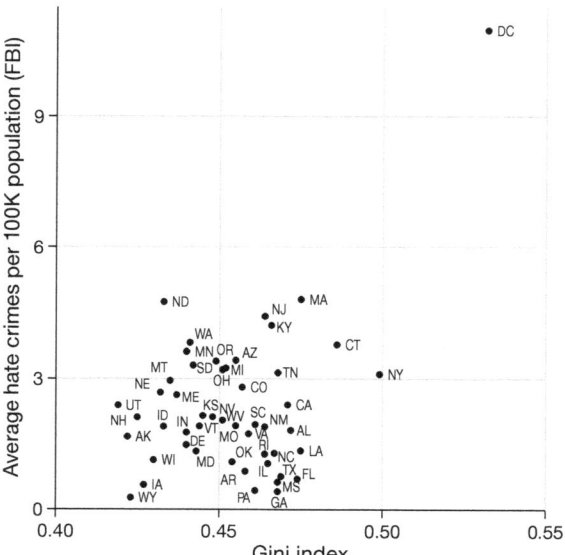

FIGURE 13.1. Plot of rates of hate crimes versus the
Gini index.

Covariance and Correlation

One way to quantify the relationship between two variables is the *covariance*. Remember that the sample variance for a single variable is computed as the average squared difference between each data point and the mean, divided by the sample size minus one:

$$s^2 = \frac{\sum_{i=1}^n (x_i - \bar{x})^2}{N - 1}$$

This tells us how far each observation is from the mean, on average, in squared units. Covariance tells us whether there is a relationship between the deviations of two different variables across observations. It is defined as

$$covariance = \frac{\sum_{i=1}^n (x_i - \bar{x})(y_i - \bar{y})}{N - 1}$$

This value is far from zero when individual data points deviate by similar amounts from their respective means on the two variables; if they are deviant in the same direction, then the covariance is positive, whereas if they are deviant in opposite directions, the covariance is negative. Let's look at a toy example first. The data are shown in table 13.1, along with their individual deviations from the mean and their cross products.

The covariance is simply the mean of the cross products, which in this case is 17.05. We don't usually use the covariance to describe relationships between variables, because it varies with the overall level of variance in the data. Instead, we would usually use the *correlation coefficient*. The correlation is computed by scaling the covariance by the standard

Table 13.1. Data for Toy Example of Covariance

x	y	y deviation	x deviation	Cross product
3	5	−3.6	−4.6	16.56
5	4	−4.6	−2.6	11.96
8	7	−1.6	0.4	−0.64
10	10	1.4	2.4	3.36
12	17	8.4	4.4	36.96

deviations of the two variables:

$$r = \frac{covariance}{s_x s_y} = \frac{\sum_{i=1}^{n}(x_i - \bar{x})(y_i - \bar{y})}{(N-1)s_x s_y}$$

In the toy example showing in table 13.1, the value is 0.89. The correlation coefficient is useful because it varies between −1 and 1 regardless of the nature of the data, which makes it easily interpretable—in fact, we already discussed the correlation coefficient when we looked at effect sizes in chapter 10. As we saw there, a correlation of 1 indicates a perfect linear relationship, a correlation of −1 indicates a perfect negative relationship, and a correlation of 0 indicates no linear relationship.

Hypothesis Testing for Correlations

The correlation value of 0.42 between hate crimes and income inequality seems to indicate a reasonably strong relationship between the two, but we can also imagine that this could occur by chance even if there is no relationship. We can test the null hypothesis that the correlation is zero, using a simple equation that lets us convert a correlation value into a t statistic:

$$t_r = \frac{r\sqrt{N-2}}{\sqrt{1-r^2}}$$

Under the null hypothesis $H_0 : r = 0$, this statistic is distributed as a t distribution with $N - 2$ degrees of freedom. We can compute this using our statistical software:

```
##
##   Pearson's product-moment correlation
##
## data:  avg_hatecrimes_per_100k_fbi and gini_index
## t = 3, df = 48, p-value = 0.002
## alternative hypothesis: true correlation is not equal to 0
## 95 percent confidence interval:
```

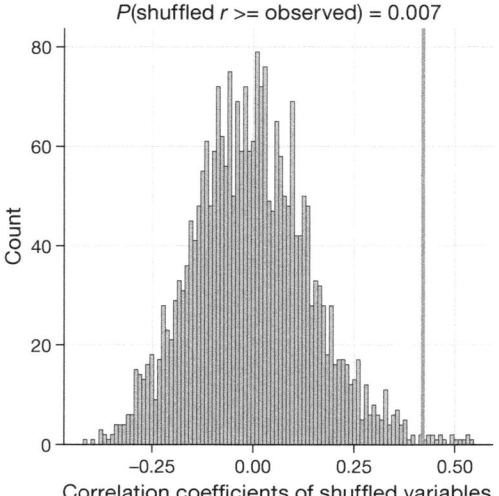

FIGURE 13.2. Histogram of correlation values under the null hypothesis, obtained by shuffling values. Observed value is denoted by the vertical line.

```
##  0.16 0.63
## sample estimates:
## cor
## 0.42
```

This test shows that the likelihood of an r value this extreme or more is quite low under the null hypothesis, so we would reject the null hypothesis of $r = 0$. Importantly, this test assumes that both variables are normally distributed. It's also worth noting that the confidence interval for the correlation is quite wide (from 0.16 to 0.63), showing that the data are consistent with a wide range of possible correlation values.

We could also test this by randomization, in which we repeatedly shuffle the values of one of the variables and compute the correlation coefficient; then we could compare our observed correlation value to this null distribution to determine how likely our observed value would be under the null hypothesis. The results are shown in figure 13.2. The p-value computed using randomization is reasonably similar to the answer given by the t test.

We could also use Bayesian inference to estimate the correlation; see the chapter appendix for more on this.

Outliers and Robust Correlations

You may have noticed something a bit odd in figure 13.1—one of the data points (the one for the District of Columbia) seems to be quite separate from the others. We refer to this

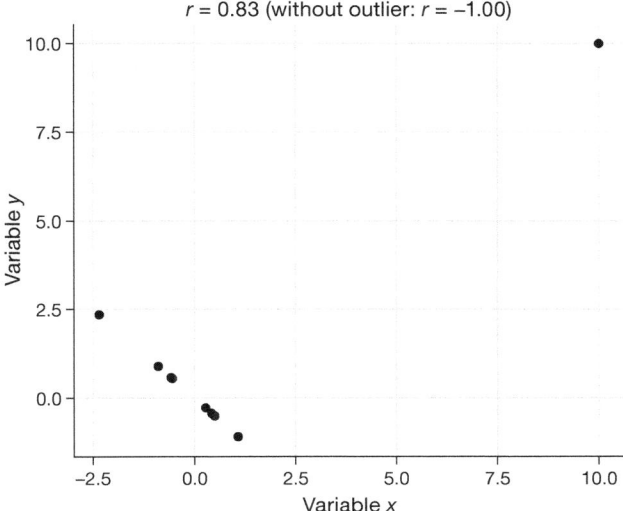

FIGURE 13.3. A simulated example of the effects of outliers on correlation. Without the outlier, the remainder of the data points have a perfect negative correlation, but the single outlier changes the correlation value to highly positive.

as an *outlier*, and the correlation coefficient is very sensitive to outliers. For example, in figure 13.3, we can see how a single outlying data point can cause a very high positive correlation value, even when the actual relationship between the other data points is perfectly negative.

One way to address outliers is to compute the correlation on the ranks of the data after ordering them, rather than on the data themselves; this is known as the *Spearman correlation*. Whereas the correlation coefficient for the example in figure 13.3 is 0.83, the Spearman correlation is -0.45, showing that the rank correlation reduces the effect of the outlier and reflects the negative relationship between the majority of the data points.

We can compute the rank correlation on the hate crime data as well:

```
##
##   Spearman's rank correlation rho
##
## data:  avg_hatecrimes_per_100k_fbi and gini_index
## S = 20146, p-value = 0.8
## alternative hypothesis: true rho is not equal to 0
## sample estimates:
##    rho
## 0.033
```

Now we see that the correlation is no longer significant (and in fact is very near zero), suggesting that the claims of the original blog post may have been incorrect due to the effect of the outlier.

Correlation and Causation

When we say that one thing *causes* another, what do we mean? There is a long history of philosophical discussion about the meaning of causality, but in statistics one way that we commonly think of causation is in terms of the effects of an experimental manipulation. That is, if we think that factor X causes factor Y, then experimentally manipulating the value of X should also change the value of Y.

In medicine, there is a set of ideas known as *Koch's postulates*, which have historically been used to determine whether a particular organism causes a disease. The basic idea is that the organism should be present in people with the disease and not present in those without it—thus, a treatment that eliminates the organism should also eliminate the disease. Further, infecting someone with the organism should cause them to contract the disease. An example of this was seen in the work of Dr. Barry Marshall, who had a hypothesis that stomach ulcers were caused by a bacterium (*Helicobacter pylori*). To demonstrate this, he infected himself with the bacterium and soon thereafter developed severe inflammation in his stomach. He then treated himself with an antibiotic, and his stomach soon recovered. He later won the Nobel Prize in Medicine for this work.

Often we would like to test causal hypotheses, but we can't actually do an experiment—either because it's impossible ("What is the relationship between human carbon emissions and the earth's climate?") or unethical ("What are the effects of severe abuse on child brain development?"). However, we can still collect data that might be relevant to those questions. For example, we can potentially collect data from children who have been abused as well as those who have not, and we can then ask whether their brain development differs.

Let's say that we did such an analysis, and we found that abused children had poorer brain development than nonabused children. Would this demonstrate that abuse *causes* poorer brain development? No. Whenever we observe a statistical association between two variables, it is certainly possible that one of those two variables causes the other. However, it is also possible that both of the variables are being influenced by a third variable; in this example, it could be that child abuse is associated with family stress, which could also cause poorer brain development through less intellectual or social engagement, food stress, or many other possible avenues. The point is that a correlation between two variables generally tells us that something is *probably* causing something else, but it doesn't tell us what is causing what.

Causal Graphs

One useful way to describe causal relationships between variables is through a *causal graph*, which shows variables as circles and the causal relationships between them as arrows. For example, figure 13.4 shows the causal relationships between study time and two variables that we think should be affected by it: exam grades and exam finishing times.

However, in reality the effects on finishing time and grades are not due directly to the amount of time spent studying, but rather to the amount of knowledge that the student gains by studying. We would usually say that knowledge is a *latent* variable—that is, we

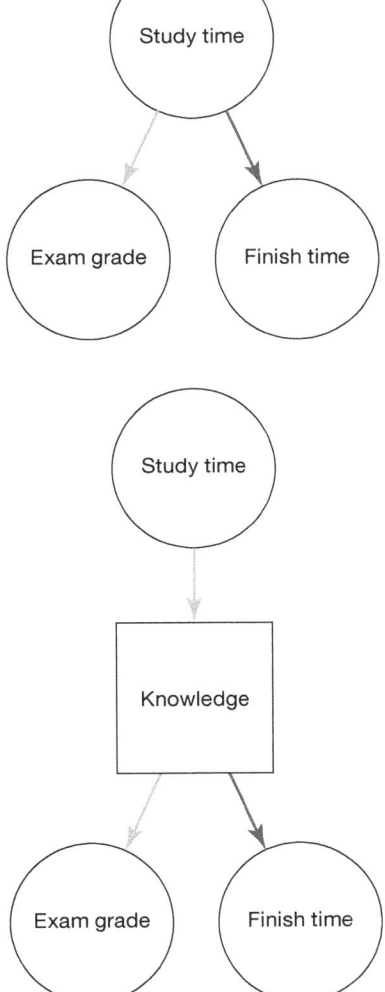

FIGURE 13.4. A graph showing causal relationships between three variables: study time, exam grades, and exam finishing time. A light gray arrow represents a positive relationship (i.e., more study time causes exam grades to increase), and a dark gray arrow represents a negative relationship (i.e., more study time causes faster completion of the exam).

FIGURE 13.5. A graph showing the same causal relationships as in figure 13.4, but also showing the latent variable (knowledge) using a square box.

can't measure it directly, but we can see it reflected in variables that we can measure (like grades and finishing times). Figure 13.5 shows this.

Here we would say that knowledge *mediates* the relationship between study time and grades/finishing times. That means that if we were able to hold knowledge constant (for example, by administering a drug that causes immediate forgetting), then the amount of study time should no longer have an effect on grades and finishing times.

Note that if we simply measured exam grades and finishing times we would generally see a negative relationship between them, because people who finish exams the fastest in general get the highest grades. However, if we were to interpret this correlation as a causal relationship, this would tell us that in order to get better grades we should actually finish the exam more quickly! This example shows how tricky the inference of causality from nonexperimental data can be.

Within statistics and machine learning, there is a very active research community that is currently studying the question of when and how we can infer causal relationships from nonexperimental data. However, these methods often require strong assumptions and must generally be used with great caution.

Suggested Reading

- *The Book of Why* by Judea Pearl. Pearl is a world leader in the development of ideas about how to understand causality from a statistical standpoint, and this book is an excellent nontechnical introduction to those ideas.

Problems

1. Which of the following statements about correlation is true? Choose all that apply.
   - The correlation is equal to the covariance scaled by the standard deviations of the two variables.
   - The correlation is equal to the covariance scaled by the variances of the two variables.
   - The correlation is equal to the covariance between Z-scored variables.
   - If the standard deviations of the variables are greater than one, then the correlation is greater than the covariance.
2. Match each figure with the most appropriate correlation value from the following list: $0, .3, .9, -0.6$.

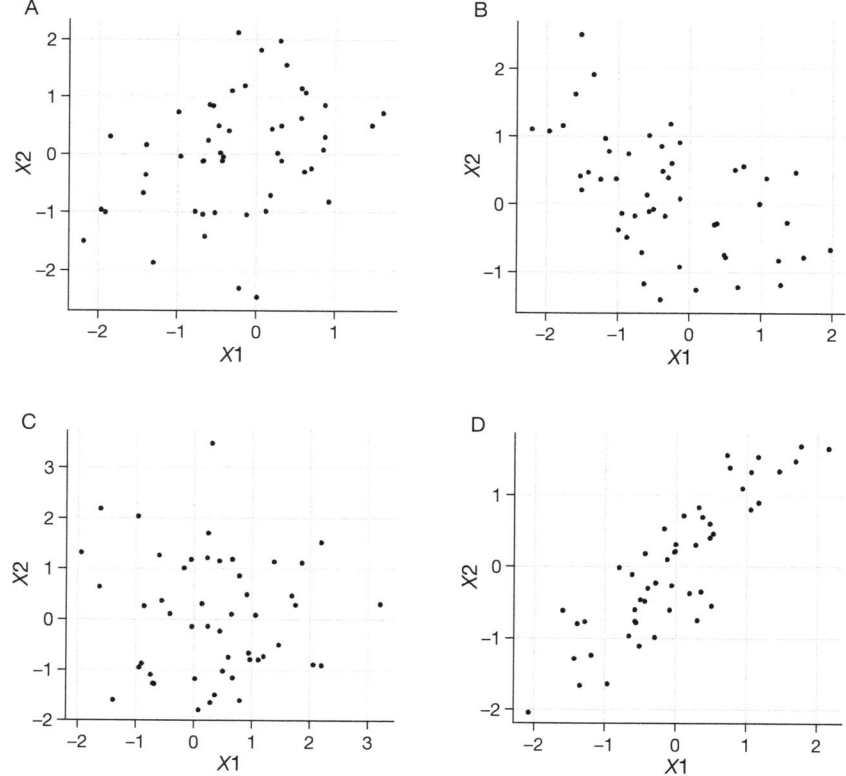

3. A medical researcher observes a negative correlation between the eating of kale and the incidence of liver cancer. What does this tell us about whether there is a causal relationship between eating kale and the developing liver cancer? Describe the different possible causal relationships that could give rise to such a result.

4. A researcher collects a set of measurements on bees to determine whether there is a relationship between the size of the colony and the perceived quality of the honey they produce. These are the data obtained for five colonies:

Size	Quality
5	37
12	44
16	61
28	77
41	82

- Using these values, compute the covariance and correlation between size and quality.
- Perform a test to determine whether the correlation between size and quality is significantly different from zero.

Appendix

Quantifying Inequality: The Gini Index

Before we look at the analysis reported in the fivethirtyeight.com story at the beginning of the chapter, it's first useful to understand how the Gini index is used to quantify inequality. The Gini index is usually defined in terms of a curve that describes the relationship between income and the proportion of the population that has income at or less than that level, known as a *Lorenz curve*. However, another way to think of it is more intuitive: it is the relative mean absolute difference between incomes, divided by two (from https://en .wikipedia.org/wiki/Gini_coefficient):

$$G = \frac{\sum_{i=1}^{n} \sum_{j=1}^{n} |x_i - x_j|}{2n \sum_{i=1}^{n} x_i}$$

Figure 13.6 shows the Lorenz curves for several different income distributions. Panel A shows an example with 10 people where everyone has exactly the same income. The length of the intervals between points are equal, indicating each person earns an identical share of the total income in the population. Panel B shows an example where income is normally distributed. Panel C shows an example with high inequality; everyone has equal income ($40,000) except for one person, who has income of $40,000,000. According to

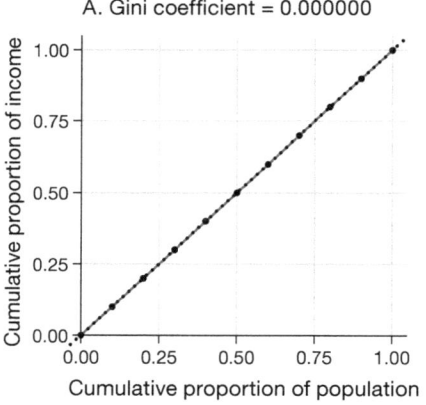

A. Gini coefficient = 0.000000

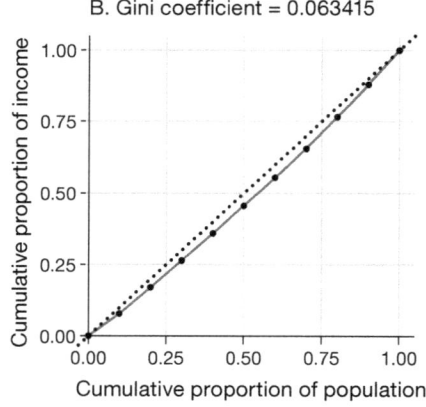

B. Gini coefficient = 0.063415

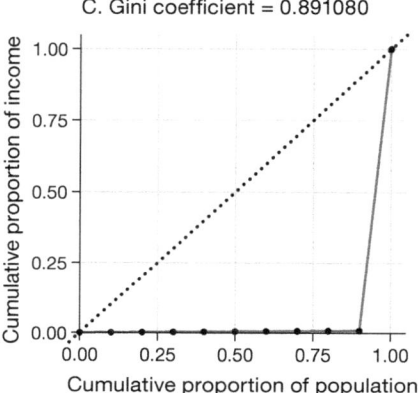

C. Gini coefficient = 0.891080

FIGURE 13.6. Lorenz curves for (A) perfect equality, (B) normally distributed income, and (C) high inequality (equal income except for one very wealthy individual).

the US Census, the United States had a Gini index of 0.469 in 2010, falling roughly halfway between our normally distributed and maximally inequal examples.

Bayesian Correlation Analysis

We can also analyze the hate crime data from the fivethirtyeight.com story using Bayesian analysis, which has two advantages. First, it provides us with a posterior probability—in this case, the probability that the correlation value exceeds zero. Second, the Bayesian estimate combines the observed evidence with a *prior*, which has the effect of *regularizing* the correlation estimate, effectively pulling it toward zero. Here we can compute it using the BayesFactor package in R.

```
## Bayes factor analysis
## --------------
## [1] Alt., r=0.333 : 21 ±0%
```

```
##
## Against denominator:
##   Null, rho = 0
## ---
## Bayes factor type: BFcorrelation, Jeffreys-beta*

## Summary of Posterior Distribution
##
## Parameter | Median |        95% CI |    BF |     Prior
## ----------------------------------------------------------
## rho       |   0.38 | [0.13, 0.59] | 20.85 | Beta (3 +- 3)
```

Notice that the correlation estimated using the Bayesian method (0.38) is slightly smaller than the one estimated using the standard correlation coefficient (0.42), because the Bayesian estimate is based on a combination of the evidence and the prior, which effectively shrinks the estimate toward zero. However, notice that the Bayesian analysis is not robust to the outlier, and it still says there is fairly strong evidence that the correlation is greater than zero (with a Bayes factor of more than 20).

14

The General Linear Model

Learning Objectives

Having read this chapter, you should be able to

- Describe the concept of linear regression and apply it to a dataset.
- Describe the concept of the general linear model and provide examples of its application.
- Describe how cross-validation can allow us to estimate the predictive performance of a model on new data.

The General Linear Model

Remember that early in the book we described the basic model of statistics:

$$data = model + error$$

where our general goal is to find the model that minimizes the error, subject to some other constraints (such as keeping the model relatively simple so that we can generalize beyond our specific dataset). In this chapter, we focus on a particular implementation of this approach, which is known as the *general linear model* (or GLM). You have already seen the general linear model in chapter 5, where we modeled height in the NHANES dataset as a function of age; here we provide a more general introduction to the concept of the GLM and its many uses. Nearly every model used in statistics can be framed in terms of the general linear model or an extension of it.

Before we discuss the general linear model, let's first define two terms that will be important for our discussion:

- *Dependent variable*: This is the outcome variable that our model aims to explain (usually referred to as Y).
- *Independent variable*: This is a variable or variables that we wish to use in order to explain the dependent variable (usually referred to as X).

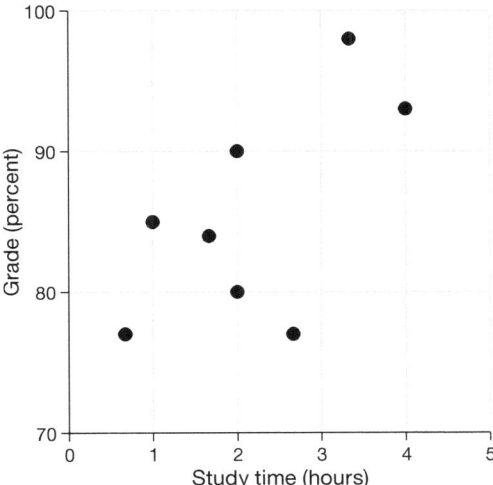

FIGURE 14.1. Relationship between study time
and grades.

Often there are multiple independent variables, though in this chapter we focus primarily on situations where there is only one independent variable in our analysis; in chapter 17, you will see examples of analyses using multiple independent variables.

In a general linear model, the dependent variable is described by a *linear combination* of independent variables that are each multiplied by a weight (which is often referred to as the Greek letter beta—β), which determines the relative contribution of that independent variable to the model prediction.

As an example, let's generate some simulated data for the relationship between study time and exam grades (figure 14.1). Given these data, we might want to engage in each of the three fundamental activities of statistics:

- *Describe*: How strong is the relationship between grades and study time?
- *Decide*: Is there a statistically significant relationship between grades and study time?
- *Predict*: Given a particular amount of study time, what grade do we expect?

In the previous chapter, we learned how to describe the relationship between two variables using the correlation coefficient. Let's use our statistical software to compute that relationship for these data and test whether the correlation is significantly different from zero:

```
##
##   Pearson's product-moment correlation
##
## data:  grade and studyTime
## t = 2, df = 6, p-value = 0.09
```

```
## alternative hypothesis: true correlation is not equal to 0
## 95 percent confidence interval:
##  -0.13  0.93
## sample estimates:
##  cor
## 0.63
```

The correlation is quite high, but notice that the confidence interval around the estimate is very wide, spanning nearly the entire range from zero to one. This is due primarily to the small sample size, which renders our statistical estimates very uncertain.

Linear Regression

We can use the general linear model to describe the relationship between two variables and to decide whether that relationship is statistically significant; in addition, the model allows us to predict the value of the dependent variable given some new value(s) of the independent variable(s). Most importantly, the general linear model allows us to build models that incorporate multiple independent variables, whereas the correlation coefficient can only describe the relationship between two individual variables.

The specific version of the GLM that we use in this case is referred to as *linear regression*. The term *regression* was coined by Francis Galton, who had noted that, when he compared parents and their children on some feature (such as height), the children of extreme parents (i.e., very tall or very short parents) generally fell closer to the mean than did their parents. (See box 14.1 for more on the complicated legacy of Galton and other early statisticians.) This is an extremely important point that we return to below.

The simplest version of the linear regression model (with a single independent variable) can be expressed as follows:

$$y = x * \beta_x + \beta_0 + \epsilon$$

The β_x value tells us how much we would expect y to change given a one-unit change in x. The intercept β_0 is an overall offset, which tells us what value we would expect y to have when $x = 0$; you may remember from our earlier modeling discussion that it is important to model the overall magnitude of the data, even if x never actually attains a value of zero. The error term ϵ refers to whatever is left over once the model has been fit; we often refer to these leftovers as the *residuals* from the model, since the term *error* can be confusing due to its multiple meanings. If we want to know how to predict y (which we call $\hat{y}$) after we estimate the β values, then we can drop the residual term:

$$\hat{y} = x * \hat{\beta}_x + \hat{\beta}_0$$

Note that this is simply the equation for a line, where $\hat{\beta}_x$ is our estimate of the slope and $\hat{\beta}_0$ is the intercept. Figure 14.2 shows an example of this model applied to the study time data.

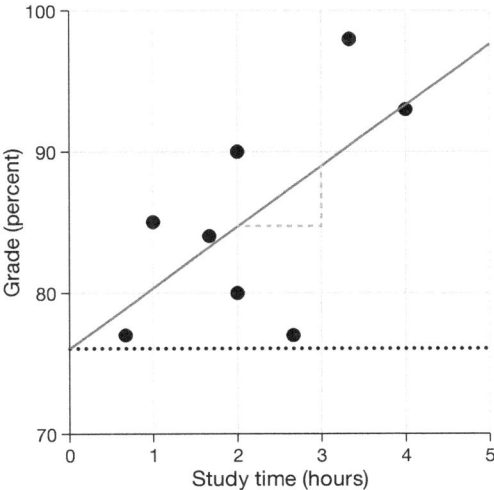

FIGURE 14.2. The linear regression solution for the study time data is shown by the solid line. The value of the intercept is equivalent to the predicted value of the *y* variable when the *x* variable is equal to zero; this is shown by the dotted line. The value of beta is equal to the slope of the line—that is, how much *y* changes for a unit change in *x*. This is shown schematically by the dashed lines, which show the degree of increase in the grade for a single unit increase in study time.

We don't go into the details of how the best-fitting slope and intercept are actually estimated from the data; if you are interested, details are available in the chapter appendix.

Box 14.1. The Complicated Legacy of Early Statisticians

Statistics, as a lens through which scientists investigate real-world questions, has always been smudged by the fingerprints of the people holding the lens.[1]

Throughout this book we mention the important work of a number of individuals who discovered key concepts of statistics, such as Francis Galton, Ronald Fisher, and Karl Pearson. While their contributions to statistics are essential, it's also important to recognize that each of these individuals held views that would now be considered repugnant and offensive. In particular, each of these individuals was involved in the development and promotion of the concept of *eugenics*, which was defined by Galton (1883, p. 25) as

the science of improving stock, which is by no means confined to questions of judicious mating, but which, especially in the case of man, takes cognisance of all influences that tend in however remote a degree to give to the more suitable races or strains of blood a better chance of prevailing speedily over the less suitable than they otherwise would have had.

1. https://nautil.us/how-eugenics-shaped-statistics-9365/.

The efforts of the eugenicists were initially focused on "positive" eugenics, which advocated for increasing the reproductive rate of high-status White individuals. However, over time the eugenics movement became primarily focused in "negative" eugenics, which promoted limitations on the fertility of individuals who were thought to be "inferior." In the early 1900s in the US, the promoters of eugenics were responsible for involuntary sterilization laws that resulted in the forced sterilization of more than 60,000 individuals (disproportionately applied to people of color) as well as restrictions on immigration. Eugenic ideas culminated in Nazi policies that ultimately resulted in the murder of more than six million individuals, primarily Jewish people but also including people of color, people with disabilities, and homosexual people. Both Fisher and Pearson expressed some degree of sympathy with the Nazi project. At his retirement dinner in 1934, Karl Pearson said the following:

> In Germany a vast experiment is in hand, and some of you may live to see its results. If it fails it will not be for want of enthusiasm, but rather because the Germans are only just starting the study of mathematical statistics in the modern sense![2]

What are we to make of this? The tools and ideas developed by Galton, Fisher, and Pearson are fundamental to statistics, and we cannot simply set them aside because they were developed by individuals who held repugnant views. What we can do is remember that, when we apply the tools of statistics, we are necessarily expressing our values in addition to our knowledge. Just as philosophers of science agree that all scientific investigation is "theory-laden" (Godfrey-Smith 2021), we must realize that we necessarily bring our values and preconceptions into any data analysis, through the data that we use (or don't use) and the questions that we choose to ask (or not ask). Users of statistics must remain aware of the potential impact that their efforts might have, particularly on those populations that have the least power to defend themselves.

2. https://profjoecain.net/karl-pearson-praised-hitler-nazi-race-hygiene/.

Regression to the Mean

The concept of *regression to the mean* was one of Galton's essential contributions to science, and it remains a critical point to understand when we interpret the results of experimental data analyses. Let's say that we want to study the effects of a reading intervention on the performance of poor readers. To test our hypothesis, we might go into a school and recruit those individuals in the bottom 25% of the distribution on some reading test, administer the intervention, and then examine their performance on the test after the intervention. Let's say that the intervention actually has no effect, such that reading scores for each individual are simply independent samples from a normal distribution. Results from a computer simulation of this hypothetical experiment are presented in table 14.1.

If we look at the difference between the mean test performance at the first and second test, it appears that the intervention helped these students substantially, as their scores

Table 14.1. Reading Scores
for Test 1 and Test 2

Test	Score
Test 1	88
Test 2	101

Note: Test 1 scores are lower because they are the basis for selecting the students. Test 2 scores are higher because they are independent from test 1.

went up by more than 10 points on the test! However, we know that, in fact, the students didn't improve at all, since in both cases the scores were simply selected from a random normal distribution. What happened was that some students scored badly on the first test simply due to random chance. If we select just those subjects on the basis of their first test scores, they are guaranteed to move back toward the mean of the entire group on the second test, even if there is no effect of training. This is the reason that any study of an intervention needs an untreated *control group* in order to interpret changes in performance due to an intervention; otherwise we are likely to be tricked by regression to the mean. In addition, the participants need to be randomly assigned to the control or treatment group so that there won't be any systematic differences between the groups (on average).

The Relationship Between Correlation and Regression

There is a close relationship between correlation coefficients and regression coefficients. Remember that the correlation coefficient is computed as the ratio of the covariance and the product of the standard deviations of x and y:

$$\hat{r} = \frac{covariance_{xy}}{s_x * s_y}$$

whereas the regression beta for x is computed as

$$\hat{\beta}_x = \frac{covariance_{xy}}{s_x * s_x}$$

Based on these two equations, we can derive the relationship between $\hat{r}$ and $\hat{\beta}$:

$$covariance_{xy} = \hat{r} * s_x * s_y$$

$$\hat{\beta}_x = \frac{\hat{r} * s_x * s_y}{s_x * s_x} = r * \frac{s_y}{s_x}$$

That is, the regression slope is equal to the correlation value multiplied by the ratio of standard deviations of y and x. One thing this tells us is that, when the standard deviations of x and y are the same (e.g., when the data have been converted to Z-scores), then the correlation estimate is equal to the regression slope estimate.

Standard Errors for Regression Models

If we want to make inferences about the regression parameter estimates, then we also need an estimate of their variability. To compute this value, we first need to compute the *residual variance* or *error variance* for the model—that is, how much variability in the dependent variable is not explained by the model. We can compute the model residuals as follows:

$$residual = y - \hat{y} = y - (x * \hat{\beta}_x + \hat{\beta}_0)$$

We then compute the sum of squared errors (SS_{error}):

$$SS_{error} = \sum_{i=1}^{n} (y_i - \hat{y}_i)^2 = \sum_{i=1}^{n} residual^2$$

and from this we compute the mean squared error (MS_{error}):

$$MS_{error} = \frac{SS_{error}}{df} = \frac{\sum_{i=1}^{n} (y_i - \hat{y}_i)^2}{N - p}$$

where the degrees of freedom (df) are determined by subtracting the number of estimated parameters (two in this case: $\hat{\beta}_x$ and $\hat{\beta}_0$) from the number of observations (N). Once we have the mean squared error, we can compute the standard error for the model as

$$SE_{model} = \sqrt{MS_{error}}$$

In order to get the standard error for a specific regression parameter estimate, SE_{β_x}, we need to rescale the standard error of the model by the square root of the sum of squares of the *x* variable:

$$SE_{\hat{\beta}_x} = \frac{SE_{model}}{\sqrt{\sum (x_i - \bar{x})^2}}$$

Statistical Tests for Regression Parameters

Once we have the parameter estimates and their standard errors, we can compute a *t* statistic to tell us the likelihood of the observed parameter estimates compared to some expected value under the null hypothesis. In this case, we test against the null hypothesis of no effect (i.e., $\beta = 0$):

$$t_{N-p} = \frac{\hat{\beta} - \beta_{expected}}{SE_{\hat{\beta}}}$$

$$t_{N-p} = \frac{\hat{\beta} - 0}{SE_{\hat{\beta}}}$$

$$t_{N-p} = \frac{\hat{\beta}}{SE_{\hat{\beta}}}$$

In general, we would use statistical software to compute these rather than computing them by hand. Here are the results from the linear model function in R:

```
##
## Call:
## lm(formula = grade ~ studyTime, data = df)
##
## Coefficients:
##              Estimate Std. Error t value Pr(>|t|)
## (Intercept)    76.16       5.16    14.76  6.1e-06 ***
## studyTime       4.31       2.14     2.01    0.091 .
## ---
## Signif. codes:  0 '***' 0.001 '**' 0.01 '*' 0.05 '.' 0.1 ' ' 1
##
## Residual standard error: 6.4 on 6 degrees of freedom
## Multiple R-squared:  0.403,   Adjusted R-squared:  0.304
## F-statistic: 4.05 on 1 and 6 DF,  p-value: 0.0907
```

In this case, we see that the intercept is significantly different from zero (which is not very interesting) and that the effect of study time on grades is marginally significant ($p = .09$)—the same p-value as the correlation test that we performed earlier.

Quantifying Goodness of Fit of the Model

Sometimes it's useful to quantify how well the model fits the data overall, and one way to do this is to ask how much of the variability in the data is accounted for by the model. This is quantified using a value called R^2 (also known as the *coefficient of determination*). If there is only one x variable, then this is easy to compute by simply squaring the correlation coefficient:

$$R^2 = r^2$$

In the case of our study time example, $R^2 = 0.4$, which means that we have accounted for about 40% of the variance in grades.

More generally, we can think of R^2 as a measure of the fraction of variance in the data that is accounted for by the model, which can be computed by breaking the variance into multiple components:

$$SS_{total} = SS_{model} + SS_{error}$$

where SS_{total} is the variance of the data (y) and SS_{model} and SS_{error} are computed as shown earlier in this chapter. Using this information, we can then compute the coefficient of determination as

$$R^2 = \frac{SS_{model}}{SS_{total}} = 1 - \frac{SS_{error}}{SS_{total}}$$

A small value of R^2 tells us that, even if the model fit is statistically significant, it may only explain a small amount of information in the data.

Fitting More Complex Models

Often we would like to understand the effects of multiple variables on some particular outcome and how they relate to one another. In the context of our study time example, let's say that we discovered that some of the students had previously taken a course on the topic. If we plot their grades (figure 14.3), we can see that those who had a prior course perform much better than those who had not, given the same amount of study time. We would like to build a statistical model that takes this into account, which we can do by extending the model that we built above:

$$\hat{y} = \hat{\beta}_1 * studyTime + \hat{\beta}_2 * priorClass + \hat{\beta}_0$$

To model whether each individual has had a previous class or not, we use what we call *dummy coding*, in which we create a new variable that has a value of one to represent

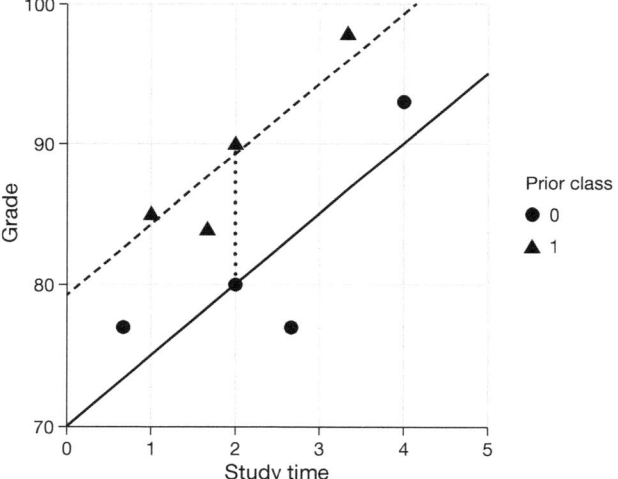

FIGURE 14.3. The relationship between study time and grades, including prior experience as an additional component in the model. The solid line relates study time to grades for students who have not had prior experience, and the dashed line relates grades to study time for students with prior experience. The dotted line corresponds to the difference in means between the two groups, which is equivalent to the regression parameter for the prior experience variable.

having had a class before and zero otherwise. This means that, for people who have had the class before, we simply add the value of $\hat{\beta}_2$ to our predicted value for them—that is, using dummy coding, $\hat{\beta}_2$ simply reflects the difference in means between the two groups. Our estimate of $\hat{\beta}_1$ reflects the regression slope over all of the data points—we are assuming that the regression slope is the same regardless of whether someone has had a class before (figure 14.3).

```
##
## Call:
## lm(formula = grade ~ studyTime + priorClass, data = df)
##
## Coefficients:
##              Estimate Std. Error t value Pr(>|t|)
## (Intercept)    70.08       3.77   18.60  8.3e-06 ***
## studyTime       5.00       1.37    3.66    0.015 *
## priorClass1     9.17       2.88    3.18    0.024 *
## ---
## Signif. codes:  0 '***' 0.001 '**' 0.01 '*' 0.05 '.' 0.1 ' ' 1
##
## Residual standard error: 4 on 5 degrees of freedom
## Multiple R-squared:  0.803,  Adjusted R-squared:  0.724
## F-statistic: 10.2 on 2 and 5 DF,  p-value: 0.0173
```

Interactions between Variables

In the previous model, we assumed that the effect of study time on grades (i.e., the regression slope) was the same for both groups. However, in some cases, we might imagine that the effect of one variable might differ depending on the value of another variable, which we refer to as an *interaction* between variables.

Let's use a new example that asks the question, What is the effect of caffeine on public speaking? First, let's generate some data and plot them. Looking at panel A of figure 14.4, there doesn't seem to be a relationship, and we can confirm that by performing linear regression on the data:

```
##
## Call:
## lm(formula = speaking ~ caffeine, data = df)
##
## Coefficients:
##              Estimate Std. Error t value Pr(>|t|)
## (Intercept)    -7.413      9.165   -0.81     0.43
## caffeine         0.158      0.151    1.11     0.28
```

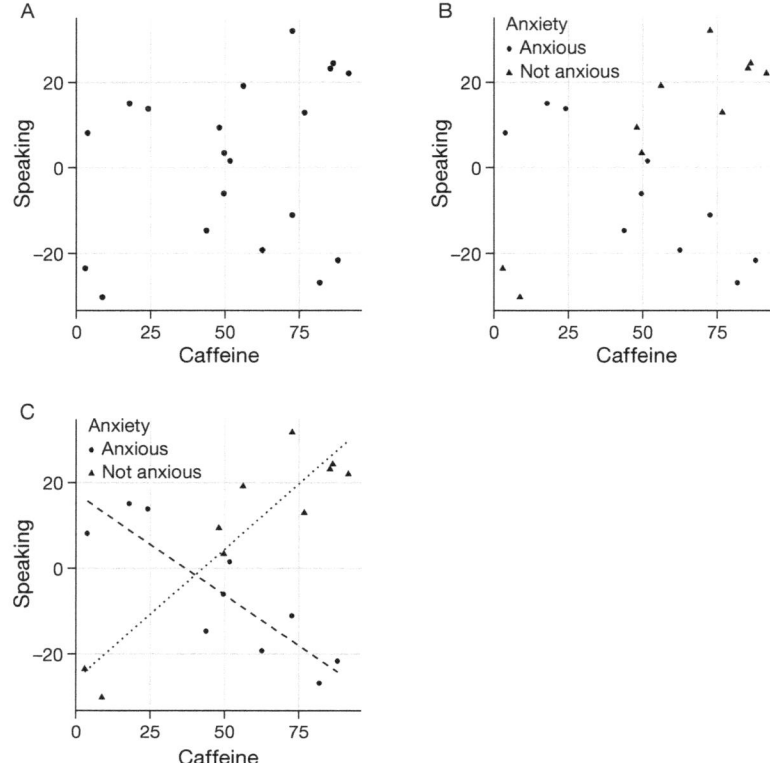

FIGURE 14.4. (A) The relationship between caffeine and public speaking. (B) The relationship between caffeine and public speaking, with anxiety represented by the shape of the data points. (C) The relationship between public speaking and caffeine, including an interaction with anxiety. This results in two lines that separately model the slope for each group (dashed for anxious, dotted for nonanxious).

```
##
## Residual standard error: 19 on 18 degrees of freedom
## Multiple R-squared:  0.0642, Adjusted R-squared:  0.0122
## F-statistic: 1.23 on 1 and 18 DF,  p-value: 0.281
```

But now let's say we find research suggesting that anxious and nonanxious people react differently to caffeine. First, let's plot the data separately for anxious and nonanxious people.

As we see from panel B in figure 14.4, it appears that the relationship between public speaking and caffeine is different for the two groups, with caffeine improving performance for people without anxiety and degrading performance for those with anxiety. We'd like to create a statistical model that addresses this question. First, let's see what happens if we just include anxiety in the model.

```
##
## Call:
## lm(formula = speaking ~ caffeine + anxiety, data = df)
##
## Coefficients:
##                    Estimate Std. Error t value Pr(>|t|)
## (Intercept)        -12.581      9.197   -1.37     0.19
## caffeine             0.131      0.145    0.91     0.38
## anxietynotAnxious   14.233      8.232    1.73     0.10
##
## Residual standard error: 18 on 17 degrees of freedom
## Multiple R-squared:  0.204,   Adjusted R-squared:  0.11
## F-statistic: 2.18 on 2 and 17 DF,  p-value: 0.144
```

Here we see there are no significant effects of either caffeine or anxiety, which might seem a bit confusing. The problem is that this model is trying to use the same slope relating public speaking to caffeine for both groups. If we want to fit them using lines with separate slopes, we need to include an *interaction* in the model, which is equivalent to fitting different lines for each of the two groups; this is often denoted by using the ∗ symbol in the model formula.

```
##
## Call:
## lm(formula = speaking ~ caffeine + anxiety + caffeine * anxiety,
##     data = df)
##
## Coefficients:
##                            Estimate Std. Error t value Pr(>|t|)
## (Intercept)                 17.4308     5.4301    3.21  0.00546 **
## caffeine                    -0.4742     0.0966   -4.91  0.00016 ***
## anxietynotAnxious          -43.4487     7.7914   -5.58  4.2e-05 ***
## caffeine:anxietynotAnxious   1.0839     0.1293    8.38  3.0e-07 ***
## ---
## Signif. codes:  0 '***' 0.001 '**' 0.01 '*' 0.05 '.' 0.1 ' ' 1
##
## Residual standard error: 8.1 on 16 degrees of freedom
## Multiple R-squared:  0.852,   Adjusted R-squared:  0.825
## F-statistic: 30.8 on 3 and 16 DF,  p-value: 7.01e-07
```

From these results, we see that there are significant effects of both caffeine and anxiety (which we call *main effects*) and an interaction between caffeine and anxiety. Panel C in figure 14.4 shows the separate regression lines for each group; the negative regression parameter for caffeine reflects the negative relationship between caffeine and performance for the anxiety group, which is treated as the baseline in the model simply because it comes first in alphabetical order. The positive regression parameter for the interaction effect (labeled `caffeine:anxietynotAnxious` above) reflects the difference in the regression slope for the nonanxious group compared to the anxious group. One important point to note is that we have to be very careful about interpreting a main effect if a significant interaction is also present, since the interaction suggests that the effect differs according to the values of another variable, and thus the overall effect is not easily interpretable.

Sometimes we want to compare the relative fit of two different models, in order to determine which is a better model; we refer to this as *model comparison*. For the models above, we can compare the goodness of fit of the model with and without the interaction, using what is called an *analysis of variance*:

```
## Analysis of Variance Table
##
## Model 1: speaking ~ caffeine + anxiety
## Model 2: speaking ~ caffeine + anxiety + caffeine * anxiety
##   Res.Df  RSS Df Sum of Sq      F Pr(>F)
## 1     17 5639
## 2     16 1046  1      4593  70.3  3e-07 ***
## ---
## Signif. codes:  0 '***' 0.001 '**' 0.01 '*' 0.05 '.' 0.1 ' ' 1
```

This tells us that there is good evidence to prefer the model with the interaction over the one without the interaction. Model comparison is relatively simple in this case because the two models are *nested*—one of the models is a simplified version of the other model, such that all the variables in the simpler model are contained in the more complex model. Model comparison with nonnested models can get much more complicated.

Beyond Linear Predictors and Outcomes

It is important to note that, despite the fact that it is called the general *linear* model, we can actually use the same machinery to model effects that don't follow a straight line (such as curves). The "linear" in the general linear model doesn't refer to the shape of the response, but instead refers to the fact that the model is linear in its parameters—that is, the predictors in the model only get multiplied by the parameters, rather than a nonlinear relationship, such as being raised to the power of a parameter. It's also common to analyze data where the outcomes are binary rather than continuous, as we saw in chapter 12. There are ways to adapt general linear models (known as *generalized* linear models) that allow this kind of analysis. We explore these models in more detail in chapter 17.

Checking Assumptions

The saying "garbage in, garbage out" is as true of statistics as anywhere else. In the case of statistical models, we have to make sure that our model is properly specified and that our data are appropriate for the model. We introduce the idea of *model diagnostics* here and delve much more deeply into them in chapter 17.

When we say that the model is "properly specified," we mean that we have included the appropriate set of independent variables in the model. We have already seen examples of misspecified models, (see figure 5.3). Remember that we saw several cases where the model failed to properly account for the data, such as failing to include an intercept. When building a model, we need to ensure that it includes all the appropriate variables.

We also need to worry about whether our model satisfies the assumptions of our statistical methods. One of the most important assumptions that we make when using the general linear model is that the residuals (that is, the difference between the model's predictions and the actual data) are normally distributed. This can fail for many reasons, either because the model is not properly specified or because the data that we are modeling are inappropriate.

We can use something called a Q-Q (quantile-quantile) plot to see whether our residuals are normally distributed. You have already encountered quantiles—they are the value that cuts off a particular proportion of a cumulative distribution. The Q-Q plot presents the quantiles of two distributions against one another; in this case, we present the quantiles of the actual data against the quantiles of a normal distribution fit to the same data. Figure 14.5 shows examples of two such Q-Q plots. Panel A shows a Q-Q plot for data from a normal distribution, while panel B shows a Q-Q plot from nonnormal data. The data points in panel B diverge substantially from the line, reflecting the fact that they are not normally distributed. We use Q-Q plots extensively in chapter 17 to check our modeling assumptions.

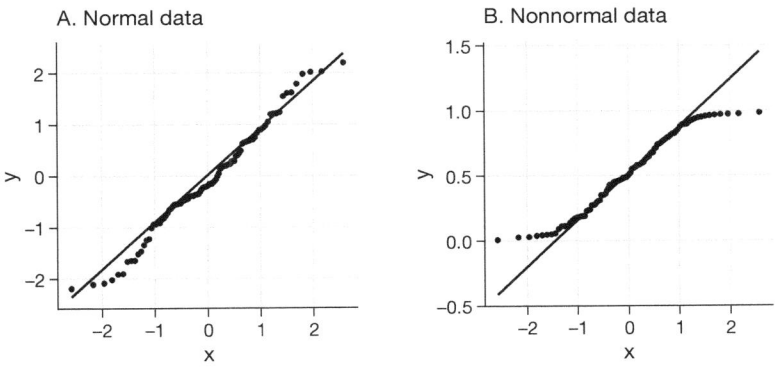

FIGURE 14.5. Q-Q plots of normal (A) and nonnormal (B) data. The line shows the point at which the x- and y-axes are equal.

What Does "Predict" Really Mean?

When we talk about "prediction" in daily life, we are generally referring to the ability to estimate the value of some variable in advance of seeing the data. However, the term is often used in the context of linear regression to refer to the fitting of a model to the data; the estimated values ($\hat{y}$) are sometimes referred to as "predictions," and the independent variables are referred to as "predictors." This has an unfortunate connotation, as it implies that our model should also be able to predict the values of new data points in the future. In reality, the fit of a model to the dataset used to obtain the parameters will nearly always be better than the fit of the model to a new dataset (Copas 1983).

As an example, let's take a sample of 48 children from the NHANES dataset and fit a regression model for weight that includes several regressors (age, height, hours spent watching TV and using the computer, and household income), along with their interactions. The root mean squared error values for this model are shown in table 14.2.

Here we see that, whereas the model on the original data shows a very good fit (only off by a few kg per individual), the same model does a much worse job of predicting the weight values for new children sampled from the same population (off by more than 25 kg per individual). This happens because the model that we specified is quite complex, since it includes not just each of the individual variables but also all possible combinations of them (i.e., their *interactions*), resulting in a model with 32 parameters. Since this is almost as many coefficients as there are data points (i.e., the heights of 48 children), the model *overfits* the data, just like the complex polynomial curve in our initial example of overfitting in chapter 5.

Another way to see the effects of overfitting is to look at what happens if we randomly shuffle the values of the weight variable (shown in the second row of the table). Randomly shuffling the value should make it impossible to predict weight from the other variables, because they should have no systematic relationship. The results in the table show that, even when there is no true relationship to be modeled (because shuffling should have obliterated the relationship), the complex model still shows a very low error in its predictions on the fitted data because it fits the noise in the specific dataset. However, when that model is applied to a new dataset, we see that the error is much larger, as it should be.

Table 14.2. Root Mean Squared Error for Model Applied to Original Data and New Data, Before and After Shuffling

Data type	RMSE (original data)	RMSE (new data)
True data	3.0	25
Shuffled data	7.8	59

Note: Shuffling the order of the y variable in essence makes the null hypothesis true.

Cross-validation

One method that has been developed to help address the problem of overfitting is known as *cross-validation*. This technique is commonly used within the field of *machine learning*, which is focused on building models that generalize well to new data, even when we don't have a new dataset to test the model. The idea behind cross-validation is that we fit our model repeatedly, each time leaving out a subset of the data, and then test the ability of the model to predict the values in each with held subset (figure 14.6).

Let's see how that would work for our weight prediction example. In this case, we will perform sixfold cross-validation, which means that we break the data into six subsets and then fit the model six times, in each case leaving out one of the subsets and then testing the model's ability to accurately predict the value of the dependent variable for those with held data points. Most statistical software provides tools to apply cross-validation to one's data. Using this function, we can run cross-validation on 100 samples from the NHANES dataset and compute the R^2 for cross-validation, along with the R^2 for the original data and a new dataset, as shown in table 14.3. In this case, R2 quantifies the amount of variance in the actual data that is accounted for by the model's predictions.

Here we see that cross-validation gives us an estimate of predictive accuracy that is much closer to what we see with a completely new dataset than it is to the inflated accuracy that we see with the original dataset. Note that using cross-validation properly is tricky, and it is recommended that you consult with an expert before using it in practice. In particular, it's essential that information about the test data doesn't "leak" into the model training procedure.

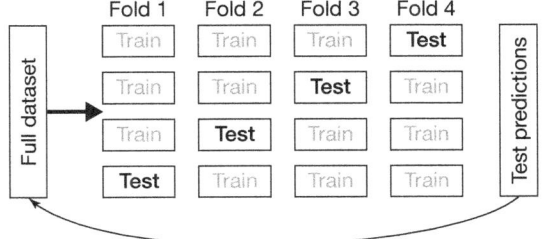

FIGURE 14.6. A schematic of the cross-validation procedure.

Table 14.3. R^2 from Cross-validation and New Data

Type of data	R^2-value
Original data	0.95
New data	0.33
Cross-validation	0.35

There are three important takeaway messages from this discussion:

- "Prediction" doesn't always mean what you think it means.
- Complex models can overfit data very badly, such that we can observe seemingly good prediction even when there is no true signal to predict.
- Claims about prediction accuracy should be viewed very skeptically unless they prediction has been made using the appropriate methods.

Suggested Reading

- *The Truth about Linear Regression*, by Cosma Shalizi. This freely available textbook from the brilliant polymath Cosma Shalizi provides a deep and opinionated introduction to the use of linear regression and its many extensions.
- *The Elements of Statistical Learning: Data Mining, Inference, and Prediction* (2nd ed.), by Trevor Hastie, Robert Tibshirani, and Jerome Friedmann. This book is widely heralded as the bible of machine learning methods, and it is freely available online. While it does involve a good bit of math, the book also takes pains to explain concepts verbally as well.

Problems

1. A researcher wants to know whether a drug improves high blood pressure. She selects a group of individuals with high blood pressure and gives them the drug, then tests their blood pressure after taking the drug for one month. She finds that their blood pressure is reduced compared to the first measurement. What concept is relevant to interpreting this outcome?
2. If the standard deviation of y is larger than the standard deviation of x, then the regresssion coefficient (beta) will be (larger/smaller) than the correlation coefficient (r).
3. The correlation is estimated by dividing the _____ by _____, whereas the regression slope is estimated by dividing _____ by _____.
4. A researcher tests whether a drug is effective at preventing flu infection using a randomized controlled trial in which participants are randomly assigned to the drug or placebo. He then follows them for one year to determine which individuals were infected with the flu.
   - What are the independent and dependent variables in this study?
   - Is regression to the mean a concern in this study?
   - Assume that the researcher finds a significant positive reduction in flu cases for the drug versus placebo groups. Can he conclude that the drug caused the difference in flu cases?
5. Which assumption may be violated if we use regular linear regression analysis when we are working with data that are part of a time series?
   - Independence
   - Normal distribution

- Equal variances
- Lack of overfitting

6. Explain in your own words the concept of an interaction in a linear model, and provide a real-world example (do not use any of the examples in the book).

7. Choose the most appropriate analysis technique for each of the following problems from these options: paired t test, independent samples t test, one-sample t test, analysis of variance.

  - Administer a measure of language skills to a group of students before and after a linguistics course, and test whether skills are better after the course compared to before.
  - Measure the blood glucose levels in a group of individuals and ask whether their average is greater than the healthy level of 100 mg/dL.
  - Compare the performance of psych majors, econ majors, and math majors on a test of world history knowledge.
  - Compare the average income of a sample of Texans and a sample of Californians.

8. Which of the following is quantified by R^2 in a linear regression model? Choose all that apply.

  - The squared residuals from the model
  - The amount of variance accounted for by the model
  - The squared correlation between the independent and dependent variable
  - The squared errors from the mean of the independent variable

9. What is the goal of cross-validation?

  - To estimate the variability of parameter estimates in the model
  - To estimate the likelihood of outliers in the data
  - To estimate the model's ability to generalize to new data
  - To estimate the statistical power of the model

Appendix

Estimating Linear Regression Parameters

We generally estimate the parameters of a linear model from data using *linear algebra*, which is the form of algebra that is applied to vectors and matrices. If you aren't familiar with linear algebra, don't worry—you won't actually need to use it here, as our software will do all the work for us. However, a brief excursion in linear algebra can provide some insight into how the model parameters are estimated in practice.

First, let's introduce the idea of vectors and matrices. A matrix is a set of numbers that are arranged in a square or rectangle, such that there are one or more *dimensions* across which the matrix varies. It is customary to place different observation units (such as people) in the rows, and different variables in the columns. Let's take our study time data from earlier in the chapter. We could arrange these numbers in a matrix, which would have eight rows (one for each student) and two columns (one for study time and one for grade). When a matrix like this is represented in software, it is often referred to as a *data frame*; this is shown in table 14.4.

Table 14.4. Study Time
Data reformatted as a Matrix

Study time	Grade
0.67	76
1.00	84
1.67	76
2.00	83
2.00	96
2.67	79
3.33	92
4.00	98

Independent variable	Design matrix			
0.67		77	1	
1.0		85	1	
1.67		84	1	
2.0	=	80	1	$* \beta + E$
2.0		90	1	
2.67		77	1	
3.33		98	1	
4.0		93	1	

FIGURE 14.7. A depiction of the linear model for the study time data in terms of matrix algebra.

We can write the general linear model in linear algebra as follows:

$$Y = X * \beta + E$$

This looks very much like the earlier equation that we used, except that the letters are all capitalized, which is meant to express the fact that they are matrices.

We know that the grade data go into the Y matrix, but what goes into the X matrix? Remember from our initial discussion of linear regression that we need to add a constant in addition to our independent variable of interest, so our X matrix (which we call the *design matrix*) needs to include two columns: one representing the study time variable and one with the same value for each individual (which we generally fill with all ones). We can view the resulting design matrix graphically (figure 14.7).

The rules of matrix multiplication tell us that the dimensions of the matrices have to match one another; in this case, the design matrix has dimensions of 8 (rows) × 2

Table 14.5. Regression Parameters
Estimated using Linear Algebra

Parameter	Estimate
Beta 1	5.2
Beta 2	74.1

(columns) and the Y variable has dimensions of 8×1. Therefore, the β matrix needs to have dimensions of 2×1, since an 8×2 matrix multiplied by a 2×1 matrix results in an 8×1 matrix (as the matching middle dimensions drop out). The interpretation of the two values in the β matrix is that they are the values to be multiplied by study time and 1, respectively, to obtain the estimated grade for each individual. We can also view the linear model as a set of equations, with one for each individual:

$$\hat{y}_1 = studyTime_1 * \beta_1 + 1 * \beta_2$$

$$\hat{y}_2 = studyTime_2 * \beta_1 + 1 * \beta_2$$

$$\dots$$

$$\hat{y}_8 = studyTime_8 * \beta_1 + 1 * \beta_2$$

Remember that our goal is to determine the best-fitting values of β given the known values of X and Y. A naive way to do this would be to solve for β using simple algebra—here we drop the error term E because it's out of our control:

$$\hat{\beta} = \frac{Y}{X}$$

The challenge here is that X and β are now matrices, not single numbers—but the rules of linear algebra tell us how to divide by a matrix, which is the same as multiplying by the *inverse* of the matrix (referred to as X^{-1}). Using our statistical software, we can apply this to the data, resulting in the estimated regression parameters shown in table 14.5.

Anyone who is interested in serious use of statistical methods is highly encouraged to invest some time into learning linear algebra, as it provides the basis for nearly all the tools that are used in standard statistics. We encounter it again in our discussion of multivariate statistics in chapter 16.

15

Comparing Means

We have already encountered a number of cases where we wanted to ask questions about the mean of a sample. In this chapter, we delve deeper into the various ways that we can compare means across groups.

Learning Objectives

Having read this chapter, you should be able to

- Describe the rationale behind the sign test.
- Describe how the *t* test can be used to compare a single mean to a hypothesized value.
- Compare the means for two paired or unpaired groups using a two-sample *t* test.
- Describe how analysis of variance can be used to test differences between more than two means.

Testing the Value of a Single Mean

The simplest question we might want to ask of a mean is whether it has a specific value. Let's say that we want to test whether the mean diastolic blood pressure value in adults from the NHANES dataset is above 80, which is the cutoff for hypertension according to the American College of Cardiology. We take a sample of 200 adults in order to ask this question; each adult has their blood pressure measured three times, and we use the average of these for our test.

One simple way to test for this difference is using the *sign test*, which asks whether the proportion of positive differences between the actual value and the hypothesized value is different than what we would expect by chance. To do this, we take the differences between each data point and the hypothesized mean value and compute their sign. If the data are normally distributed and the actual mean is equal to the hypothesized mean, then the proportion of values above (or below) the hypothesized mean should be 0.5, such that the proportion of positive differences should also be 0.5. In our sample, we see that 19.0% of individuals have a diastolic blood pressure above 80. We can then use a *binomial test* to ask whether this proportion of positive differences is greater than 0.5; this test simply uses the binomial distribution (which we have already encountered) to determine the probability of a value at least as extreme as the one we have observed given a particular null hypothesis (in this case, p ≤ 0.5):

```
##
##   Exact binomial test
##
## data:   npos and nrow(NHANES_sample)
## number of successes = 38, number of trials = 200, p-value = 1
## alternative hypothesis: true probability of success is
## greater than 0.5
## sample estimates:
## probability of success
##                    0.19
```

Here we see that the proportion of individuals with positive signs is not very surprising under the null hypothesis of p ≤ 0.5, which is expected given that the observed proportion of people with blood pressure above 80 is actually less than 0.5.

We can also ask this question using Student's *t* test, which we discussed in chapter 9. We refer to the mean as $\bar{X}$ and the hypothesized population mean as μ. Then the t-test for a single mean is

$$t = \frac{\bar{X} - \mu}{SEM}$$

where *SEM* (as you may remember from chapter 7) is defined as

$$SEM = \frac{\hat{\sigma}}{\sqrt{n}}$$

In essence, the *t* statistic asks how large the deviation of the sample mean from the hypothesized quantity is with respect to the sampling variability of the mean.

We can compute this for the NHANES dataset using our statistical software:

```
##
##   One-Sample t-test
##
## data:   BPDiaAve
## t = -55, df = 4593, p-value = 1
## alternative hypothesis: true mean is greater than 80
## sample estimates:
## mean of x
##        70
```

This shows us that the mean diastolic blood pressure in the dataset (69.5) is actually much lower than 80, so our test for whether it is above 80 is very far from significance.

Remember that a large p-value doesn't provide us with evidence in favor of the null hypothesis, since we had already assumed that the null hypothesis was true to start with. However, as we discussed in chapter 11 we can use the Bayes factor to quantify evidence for or against the null hypothesis:

```
## Bayes factor analysis
## --------------
## [1] Alt., r=0.707 -Inf<d<80     : 2.7e+16  ±NA%
## [2] Alt., r=0.707 !(-Inf<d<80)  : NaNe-Inf ±NA%
##
## Against denominator:
##   Null, mu = 80
## ---
## Bayes factor type: BFoneSample, JZS
```

The first Bayes factor listed here $(2.73 * 10^{16})$ denotes the fact that there is exceedingly strong evidence in favor of the null hypothesis over the alternative.

Comparing Two Means

A more common question that often arises in statistics is whether there is a difference between the means of two different groups. Let's say that we would like to know whether regular marijuana smokers watch more television, which we can also ask using the NHANES dataset. We take a sample of 200 individuals from the dataset and test whether the number of hours of television watching per day is related to regular marijuana use. Panel A of figure 15.1 shows these data using a violin plot.

We can also use Student's t-test to test for differences between two groups of independent observations (as we saw in chapter 9); later in the chapter we turn to cases where the observations are not independent. As a reminder, the t statistic for comparison of two independent groups is computed as

$$t = \frac{\bar{X}_1 - \bar{X}_2}{\sqrt{\frac{S_1^2}{n_1} + \frac{S_2^2}{n_2}}}$$

where $\bar{X}_1$ and $\bar{X}_2$ are the means of the two groups, S_1^2 and S_2^2 are the variances for each of the groups, and n_1 and n_2 are the sizes of the two groups. Under the null hypothesis of no difference between means, this statistic is distributed according to a t distribution, with degrees of freedom computed using the Welch test (as discussed previously) since the number of individuals differs between the two groups. In this case, we started with the specific hypothesis that smoking marijuana is associated with greater TV watching, so we will use a one-tailed test. Here are the results from our statistical software:

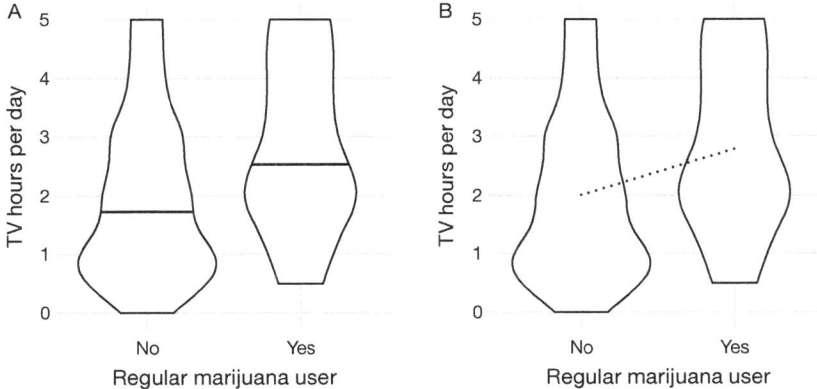

FIGURE 15.1. (A) Violin plot showing distributions of TV watching separated by regular marijuana use. (B) Violin plots showing data for each group, with a dotted line connecting the predicted values for each group, computed on the basis of the results of the linear model.

```
##
##   Welch Two-Sample t-test
##
## data:   TVHrsNum by RegularMarij
## t = -3, df = 85, p-value = 6e-04
## alternative hypothesis: true difference in means between
## group No and group Yes is less than 0
## sample estimates:
##   mean in group No    mean in group Yes
##   2.0                 2.8
```

In this case, we see that there is a statistically significant difference between groups, in the expected direction: regular pot smokers watch more TV.

The *t* test as a linear model

The *t* test is often presented as a specialized tool for comparing means, but it can also be viewed as an application of the general linear model. In this case, the model would look like this:

$$\hat{y} = \hat{\beta}_1 * marijuana + \hat{\beta}_0$$

Since marijuana use is a binary variable, we treat it as a *dummy variable*, as discussed in the previous chapter, setting it to a value of one for smokers and zero for nonsmokers. In that case, $\hat{\beta}_1$ is simply the difference in means between the two groups, and $\hat{\beta}_0$ is the mean for the group that is coded as zero. We can fit this model using the general linear model function in our statistical software, and see that it gives the same *t* statistic as the *t* test above, except that it's positive in this case because of the way our software arranges the groups:

```
##
## Call:
## lm(formula = TVHrsNum ~ RegularMarij, data = NHANES_sample)
##
## Coefficients:
##                  Estimate Std. Error t value Pr(>|t|)
## (Intercept)         2.007      0.116   17.27  < 2e-16 ***
## RegularMarijYes     0.778      0.230    3.38  0.00087 ***
## ---
## Signif. codes:  0 '***' 0.001 '**' 0.01 '*' 0.05 '.' 0.1 ' ' 1
##
## Residual standard error: 1.4 on 198 degrees of freedom
## Multiple R-squared:  0.0546, Adjusted R-squared:  0.0498
## F-statistic: 11.4 on 1 and 198 DF,  p-value: 0.000872
```

We can also view the linear model results graphically (panel B of figure 15.1). In this case, the predicted value for nonsmokers is $\hat{\beta}_0$ (2.0), and the predicted value for smokers is $\hat{\beta}_0 + \hat{\beta}_1$ (2.8).

To compute the standard errors for this analysis, we can use exactly the same equations that we used for linear regression—since this really is just another example of linear regression. In fact, if you compare the p-value from the *t* test above with the p-value in the linear regression analysis for the marijuana use variable, you will see that the one from the linear regression analysis is exactly twice the one from the *t* test, because the linear regression analysis is performing a two-tailed test.

Effect Sizes for Comparing Two Means

The most commonly used effect size for a comparison between two means is Cohen's d, which (as you may remember from chapter 10) is an expression of the effect size in terms of standard deviation units. For the *t* test estimated using the general linear model outlined in the preceding section (i.e., with a single dummy-coded variable), this is expressed as

$$d = \frac{\hat{\beta}_1}{\sigma_{residual}}$$

We can obtain these values from the linear model analysis output above, giving us $d = 0.55$, which we would generally interpret as a medium-sized effect.

We can also compute R^2 for this analysis, which tells us how what proportion of the variance in TV watching is accounted for by marijuana smoking. This value (which is reported at the bottom of the summary of the linear model analysis above) is 0.05, which tells us that while the effect may be statistically significant, it accounts for relatively little of the variance in TV watching.

Bayes Factor for Mean Differences

As we discussed in chapter 11, Bayes factors provide a way to better quantify evidence in favor of or against the null hypothesis of no difference. We can perform that analysis on the same data:

```
## Bayes factor analysis
## --------------
## [1] Alt., r=0.707 0<d<Inf     : 0.041 ±0%
## [2] Alt., r=0.707 !(0<d<Inf) : 61    ±0%
##
## Against denominator:
##   Null, mu1-mu2 = 0
## ---
## Bayes factor type: BFindepSample, JZS
```

Because of the way the data are organized, the second line shows us the relevant Bayes factor for this analysis, which is 61.4. This shows us that the evidence against the null hypothesis is quite strong.

Comparing Paired Observations

In experimental research, we often use *within-subjects* designs, in which we compare the same person on multiple measurements. The measurements that come from this kind of design are often referred to as *repeated measures*. For example, in the NHANES dataset, blood pressure was measured three times. Let's say that we are interested in testing whether there is a difference in mean systolic blood pressure between the first and second measurement across individuals in our sample (figure 15.2).

We see that there does not seem to be much of a difference in mean blood pressure (about one point) between the first and second measurement. First, let's test for a difference using an independent samples t test, which ignores the fact that pairs of data points come from the same individuals.

```
##
##   Two-Sample t-test
##
## data:  BPsys by time point
## t = 0.6, df = 398, p-value = 0.5
## alternative hypothesis: true difference in means between
## group BPsys1 and group BPsys2 is not equal to 0
```

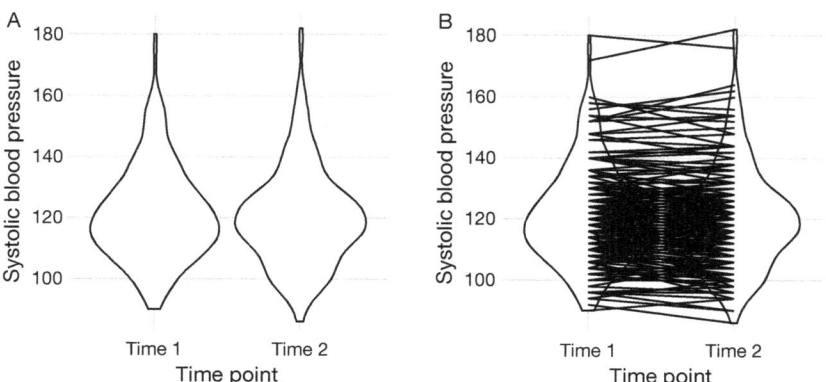

FIGURE 15.2. (A) Violin plot of systolic blood pressure on first and second recording. (B) Same violin plot with lines connecting the two data points for each individual.

```
## sample estimates:
## mean in group BPsys1    mean in group BPsys2
## 121                     120
```

This analysis shows no significant difference. However, this analysis is inappropriate since it assumes that the two samples are independent, when in fact they are not, since the data come from the same individuals. We can plot the data with a line for each individual to show this (panel B in figure 15.2).

In this analysis, what we really care about is whether the blood pressure for each person changed in a systematic way between the two measurements. So another way to represent the data is to compute the difference between the two time points for each individual and then analyze the difference scores, rather than analyzing the individual measurements. In figure 15.3, we show a histogram of these difference scores, with the vertical line denoting the mean difference.

Sign Test

As we discussed at the beginning of the chapter, one simple way to test for differences is using the *sign test*. To do this, we take the differences and compute their sign, and then we use a binomial test to ask whether the proportion of positive signs differs from 0.5.

```
##
##   Exact binomial test
##
## data:  npos and nrow(NHANES_sample)
## number of successes = 96, number of trials = 200, p-value = 0.6
## alternative hypothesis: true probability of success is not equal to 0.5
## sample estimates:
## probability of success
##                  0.48
```

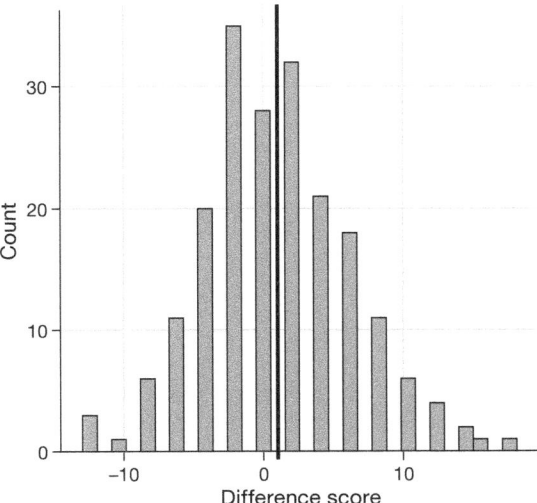

FIGURE 15.3. Histogram of difference scores between the first and second blood pressure measurement. The vertical line represents the mean difference in the sample.

Here we see that the proportion of individuals with positive signs (0.48) is not large enough to be surprising under the null hypothesis of $p = 0.5$. However, one problem with the sign test is that it is throwing away information about the magnitude of the differences and thus might be missing something.

Paired t Test

A more common strategy for comparing two observations is to use a *paired t test*, which is equivalent to a one-sample *t* test for whether the mean difference between the measurements within each person is zero. We can compute this using our statistical software, telling it that the data points are paired:

```
##
##  Paired t-test
##
## data:  BPsys by time point
## t = 3, df = 199, p-value = 0.007
## alternative hypothesis: true mean difference is not equal to 0
## 95 percent confidence interval:
##   0.29 1.75
## sample estimates:
## mean difference
##               1
```

With this analysis, we see that there is in fact a significant difference between the two measurements. Let's compute the Bayes factor to see how much evidence is provided by the result:

```
## Bayes factor analysis
## --------------
## [1] Alt., r=0.707 : 3 ±0.01%
##
## Against denominator:
##    Null, mu = 0
## ---
## Bayes factor type: BFoneSample, JZS
```

The observed Bayes factor of 3 tells us that, although the effect was significant in the paired *t* test, it actually provides very weak evidence in favor of the alternative hypothesis.

The paired *t* test can also be defined in terms of a linear model, using something called a *mixed-effects model*, which we discuss in more detail in chapter 17.

Comparing More Than Two Means

Often we want to compare more than two means to determine whether any of them differ from one another. Let's say that we are analyzing data from a clinical trial for the treatment of high blood pressure. In the study, volunteers are randomized to one of three conditions: drug 1, drug 2, or placebo. Let's generate some data and plot them (figure 15.4).

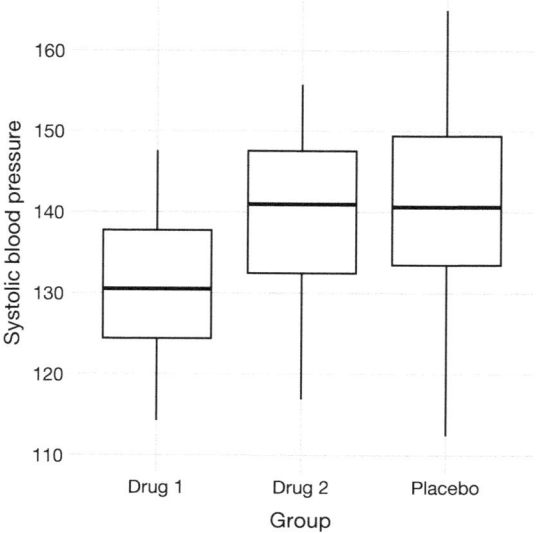

FIGURE 15.4. Box plots showing blood pressure for three different groups in our clinical trial.

Analysis of Variance

We would first like to test the null hypothesis that the means of all the groups are equal— that is, neither of the treatments had any effect compared to the placebo. We can do this using a method called *analysis of variance* (ANOVA). This is one of the most commonly used methods in psychological statistics; we only scratch the surface here. The basic idea behind ANOVA is one that we already discussed in chapter 14 on the general linear model, and in fact ANOVA is just a label for a specific version of such a model in which there are multiple dummy variables.

Remember from the last chapter that we can partition the total variance in the data (SS_{total}) into the variance that is explained by the model (SS_{model}) and the variance that is not (SS_{error}). We can then compute a *mean square* for each of these by dividing them by their degrees of freedom; for the error this is $N - p$ (where p is the number of means that we have computed), and for the model this is $p - 1$:

$$MS_{model} = \frac{SS_{model}}{df_{model}} = \frac{SS_{model}}{p - 1}$$

$$MS_{error} = \frac{SS_{error}}{df_{error}} = \frac{SS_{error}}{N - p}$$

With ANOVA, we want to test whether the variance accounted for by the model is greater than what we would expect by chance, under the null hypothesis of no differences between means. Whereas for the t distribution the expected value is zero under the null hypothesis, that's not the case here, since sums of squares are always positive numbers. Fortunately, there is another theoretical distribution that describes how ratios of sums of squares are distributed under the null hypothesis: the F distribution (figure 15.5). This distribution has two degrees of freedom, which correspond to the degrees of freedom for the numerator (which in this case is the model) and the denominator (which in this case is the error).

To create an ANOVA model, we extend the idea of dummy coding that you encountered earlier in the chapter. Remember that for the t test comparing two means we created a single dummy variable that took the value of one for one of the conditions and zero for the others. Here we extend that idea by creating two dummy variables—one that codes for the drug 1 condition and the other that codes for the drug 2 condition. Just as in the t test, we will have one condition (in this case, the placebo) that doesn't have a dummy variable, and thus represents the baseline against which the others are compared; its mean defines the intercept of the model. Using dummy coding for drugs 1 and 2, we can fit a model using the same approach that we used previously:

```
##
## Call:
## lm(formula = sysBP ~ d1 + d2, data = df)
```

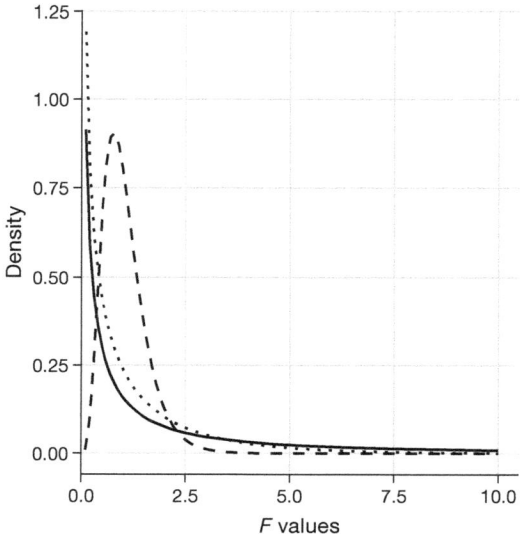

FIGURE 15.5. *F* distributions under the null hypothesis, for different values of degrees of freedom.

```
##
## Coefficients:
##              Estimate Std. Error t value Pr(>|t|)
## (Intercept)    141.60       1.66   85.50  < 2e-16 ***
## d1             -10.24       2.34   -4.37  2.9e-05 ***
## d2              -2.03       2.34   -0.87     0.39
## ---
## Signif. codes:  0 '***' 0.001 '**' 0.01 '*' 0.05 '.' 0.1 ' ' 1
##
## Residual standard error: 9.9 on 105 degrees of freedom
## Multiple R-squared:  0.169,  Adjusted R-squared:  0.154
## F-statistic: 10.7 on 2 and 105 DF,  p-value: 5.83e-05
```

The output from this command provides us with two things. First, it shows us the result of a *t* test for each of the dummy variables, which basically tell us whether each of the conditions separately differs from the placebo; it appears that drug 1 does whereas drug 2 does not. However, if we wanted to interpret these results, we would need to correct the p-values to account for the fact that we have done multiple hypothesis tests; we will see an example of how to do this in chapter 17.

Remember that the hypothesis we started out wanting to test was whether there was any difference between any of the conditions; we refer to this as an *omnibus* hypothesis test, and it is the test that is provided by the *F* statistic. The *F* statistic basically tells us whether our model is better than a simple model that just includes an intercept. In this case, we see that the F test is highly significant, consistent with our impression that there did seem to be differences between the groups (which in fact we know there were, because we created the data).

Problems

1. The local police would like to know whether the average speed of drivers on a road near the elementary school is greater than the posted speed limit of 25 miles per hour. They perform a test in which they measure the speed of 20 cars, which had a mean speed of 32 miles per hour with a standard deviation of 5. Using these values, perform a one-sample t test to determine whether the mean speed is greater than 25.
2. In order to compare the means of two groups using a linear model, we need to create what kind of variable?
   - Dummy variable
   - Interaction variable
   - Polynomial variable
   - Standardized variable
3. The paired t test is equivalent to
   - A one-sample t test for the average of the paired observations.
   - A one-sample t test for the difference between the paired observations.
   - A two-sample t test for independent observations.
   - An analysis of variance for two independent groups.
4. Analysis of variance is an appropriate method for
   - Determining the linear relationship between two continuous variables.
   - Testing the mean of a single group versus zero.
   - Identifying differences in variance between groups.
   - Testing for differences between three or more means.

16

Multivariate Statistics

The term *multivariate* refers to analyses that involve more than one random variable. Whereas we have seen examples where the model included multiple variables (as in linear regression), in those cases we were specifically interested in how the variation in a *dependent variable* could be explained in terms of one or more *independent variables* that are usually specified by the experimenter rather than measured. In a multivariate analysis, we generally treat all of the variables as equals, and seek to understand how they relate to one another as a group.

Learning Objectives

Having read this chapter, you should be able to

- Describe the distinction between supervised and unsupervised learning.
- Employ visualization techniques, including heatmaps, to visualize the structure of multivariate data.
- Understand the concept of clustering and how it can be used to identify structure in data.
- Understand the concept of dimensionality reduction.
- Describe how principal component analysis and factor analysis can be used to perform dimensionality reduction.

Varieties of Multivariate Analysis

There are many different kinds of multivariate analysis, but we focus on two major approaches in this chapter. First, we may simply want to understand and visualize the structure that exists in the data, by which we usually mean which variables or observations are related. We would usually define *related* in terms of some measure that indexes the distance between the values across variables. One important method that fits under this category is known as *clustering*, which aims to find clusters of variables or observations that are similar across variables.

Second, we may want to take a large number of variables and reduce them to a smaller number of variables in a way that retains as much information as possible. This is referred to as *dimensionality reduction*, where *dimensionality* refers to the number of variables in the dataset. We discuss two techniques commonly used for dimensionality reduction: *principal component analysis* and *factor analysis*.

Clustering and dimensionality reduction are often classified as forms of *unsupervised learning*; this is in contrast to *supervised learning*, which characterizes models such as linear regression that you've already learned about. The reason that we consider linear regression to be "supervised" is that we know the value of the thing we are trying to predict (i.e., the dependent variable), and we are trying to find the model that best predicts those values. In unsupervised learning, we don't have a specific value that we are trying to predict; instead, we are trying to discover structure in the data that might be useful for understanding what's going on, which generally requires some assumptions about what kind of structure we want to find.

One thing you will discover in this chapter is that, whereas there is generally a "right" answer in supervised learning (once we have agreed upon how to determine the "best" model, such as the sum of squared errors), often there is not an agreed-upon "right" answer in unsupervised learning. Different unsupervised learning methods can give very different answers about the same data, and there is usually no way in principle to determine which of these is "correct," as it depends on the goals of the analysis and the assumptions that one is willing to make about the mechanisms that give rise to the data. Some people find this frustrating, while others find it exhilarating; it will be up to you to figure out which of these camps you fall into.

Multivariate Data: An Example

As an example of multivariate analysis, we look at a dataset collected by my group and published in Eisenberg et al. 2019. This dataset is useful both because it has a large number of interesting variables collected on a relatively large number of individuals and because it is freely available online, so that you can explore it further on your own.

This study was performed because we were interested in understanding how several different aspects of psychological function are related to one another, focusing particularly on psychological measures of self-control and related concepts. Participants performed a 10-hour-long battery of cognitive tests and surveys over the course of a week. In this first example, we focus on variables related to two specific aspects of self-control; you will see another analysis of these data in chapter 17. *Response inhibition* is defined as the ability to quickly stop an action, and in this study it was measured using a set of tasks known as *stop-signal tasks*. The variable of interest for these tasks is an estimate of how long it takes individuals to stop themselves, known as the *stop-signal reaction time* (SSRT), of which there are four different measures in the dataset. *Impulsivity* is defined as the tendency to make decisions on impulse, without regard to potential consequences and longer-term goals. The study includes a number of different surveys measuring impulsivity, but we focus on the UPPS-P survey, which assesses five different facets of impulsivity.

After these scores are computed for each of the 522 participants in Eisenberg's study, we end up with nine numbers for each individual. We treat each of these variables as a *dimension* of the dataset; while multivariate data can sometimes have thousands or even millions of dimensions, it's useful to see how the methods work with a small number of dimensions first.

Visualizing Multivariate Data

A fundamental challenge of multivariate data is that the human eye and brain are simply not equipped to visualize data with more than three dimensions. There are various tools that we can use to try to visualize multivariate data, but all of them break down as the number of variables grows. Once the number of variables becomes too large to directly visualize, the most fruitful approach is usually to first reduce the number of dimensions (as discussed further below) and then visualize that reduced dataset.

Scatterplot of Matrices

One useful way to visualize a small number of variables is to plot each pair of variables against one another, sometimes known as a *scatterplot of matrices*; an example is shown in figure 16.1. Each row/column in the figure refers to a single variable—in this case, one of our psychological variables from the self-control dataset example discussed earlier. The diagonal elements on the plot show the distribution of each variable as a histogram. The elements below the diagonal show a scatterplot for each pair of matrices, overlaid with a regression line describing the relationship between the variables. The elements above the diagonal show the correlation coefficient for each pair of variables. When the number of variables is relatively small (about 10 or less) this can be a useful way to get good insight into a multivariate dataset. What you can immediately see is that the correlations are high between each of the SSRT variables and between each of the UPPS impulsivity variables, but the correlations between those two types of variables are all very low. This is our first inkling that there are two sets of related variables in this dataset.

Heatmap

In some cases, we wish to visualize the relationships between a large number of variables at once, usually focusing on the correlation coefficient. A useful way to do this can be to plot the correlation values as a *heatmap*, in which the color of the map relates to the value of the correlation. Figure 16.2 shows an example with a relatively small number of variables, using our self-control data from above. In this case, we see that there are strong intercorrelations within the SSRT variables and within the UPPS variables, with relatively little correlation between the two sets of variables.

Heatmaps become particularly useful for visualizing correlations between large numbers of variables. We can use brain imaging data as an example. It is common for

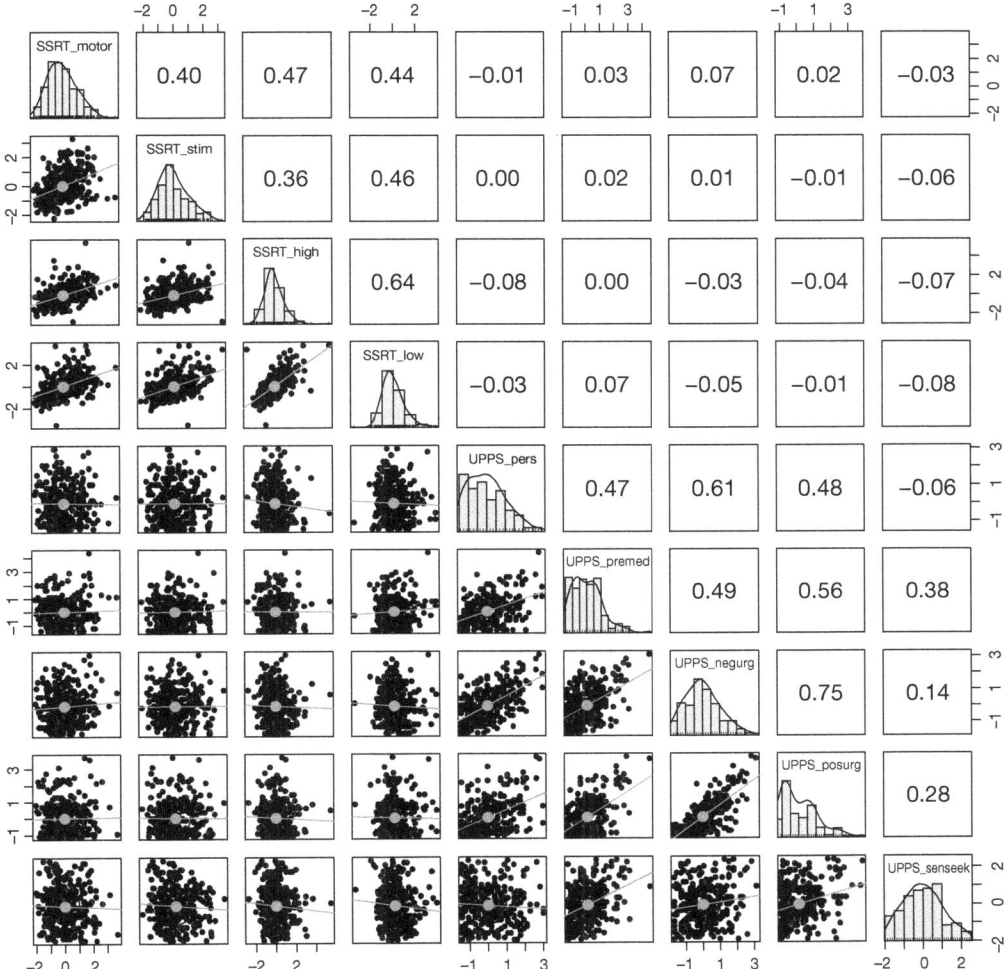

FIGURE 16.1. Scatterplot of matrices for the nine variables in the self-control dataset. The diagonal elements in the matrix show the histogram for each of the individual variables. The images to the lower left show scatterplots of the relationship between each pair of variables, and the numbers to the upper right show the correlation coefficients for each pair of variables.

neuroscience researchers to collect data about brain function from a large number of locations in the brain using functional magnetic resonance imaging (fMRI), and then to assess the correlation between those locations to measure "functional connectivity" between the regions. For example, figure 16.3 shows a heatmap for a large correlation matrix, based on activity in over 300 regions in the brain of a single individual (yours truly). The presence of clear structure in the data pops out simply by looking at the heatmap. In particular, we see large sets of brain regions whose activity is highly correlated with each other (visible in the light gray blocks along the diagonal of the correlation matrix), whereas these blocks are also strongly negatively correlated with other blocks (visible in the dark gray

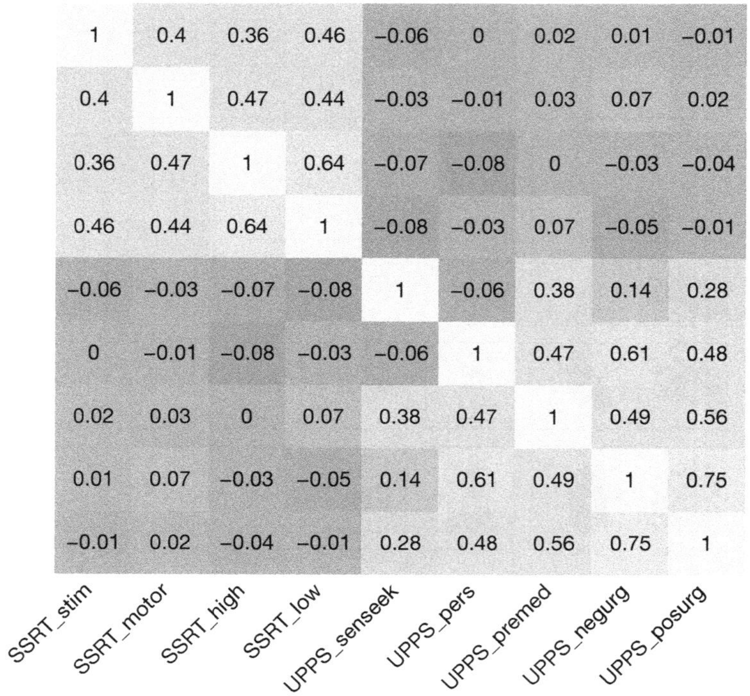

1	0.4	0.36	0.46	−0.06	0	0.02	0.01	−0.01
0.4	1	0.47	0.44	−0.03	−0.01	0.03	0.07	0.02
0.36	0.47	1	0.64	−0.07	−0.08	0	−0.03	−0.04
0.46	0.44	0.64	1	−0.08	−0.03	0.07	−0.05	−0.01
−0.06	−0.03	−0.07	−0.08	1	−0.06	0.38	0.14	0.28
0	−0.01	−0.08	−0.03	−0.06	1	0.47	0.61	0.48
0.02	0.03	0	0.07	0.38	0.47	1	0.49	0.56
0.01	0.07	−0.03	−0.05	0.14	0.61	0.49	1	0.75
−0.01	0.02	−0.04	−0.01	0.28	0.48	0.56	0.75	1

SSRT_stim, SSRT_motor, SSRT_high, SSRT_low, UPPS_senseek, UPPS_pers, UPPS_premed, UPPS_negurg, UPPS_posurg

FIGURE 16.2. Heatmap of the correlation matrix for the nine self-control variables. The light gray blocks along the diagonal (top left to bottom right) highlight the higher correlations within the two subsets of variables.

blocks seen off the diagonal). Heatmaps are a powerful tool for easily visualizing large data matrices.

Clustering

Clustering refers to a set of methods for identifying groups of related observations or variables within a dataset, based on the similarity of the values of the observations. Usually, this similarity is quantified in terms of some measure of the *distance* between multivariate values. The clustering method then finds the set of groups that have the lowest distance between their members.

One commonly used measure of distance for clustering is the *Euclidean distance*, which is basically the length of the line that connects two data points. Figure 16.4 shows an example of a dataset with two data points and two dimensions (x and y). The Euclidean distance between these two points is the length of the dotted line that connects the points in space.

The Euclidean distance is computed by squaring the differences in the locations of the points in each dimension, adding these squared differences, and then taking the square root. When there are two dimensions x and y, this would be computed as

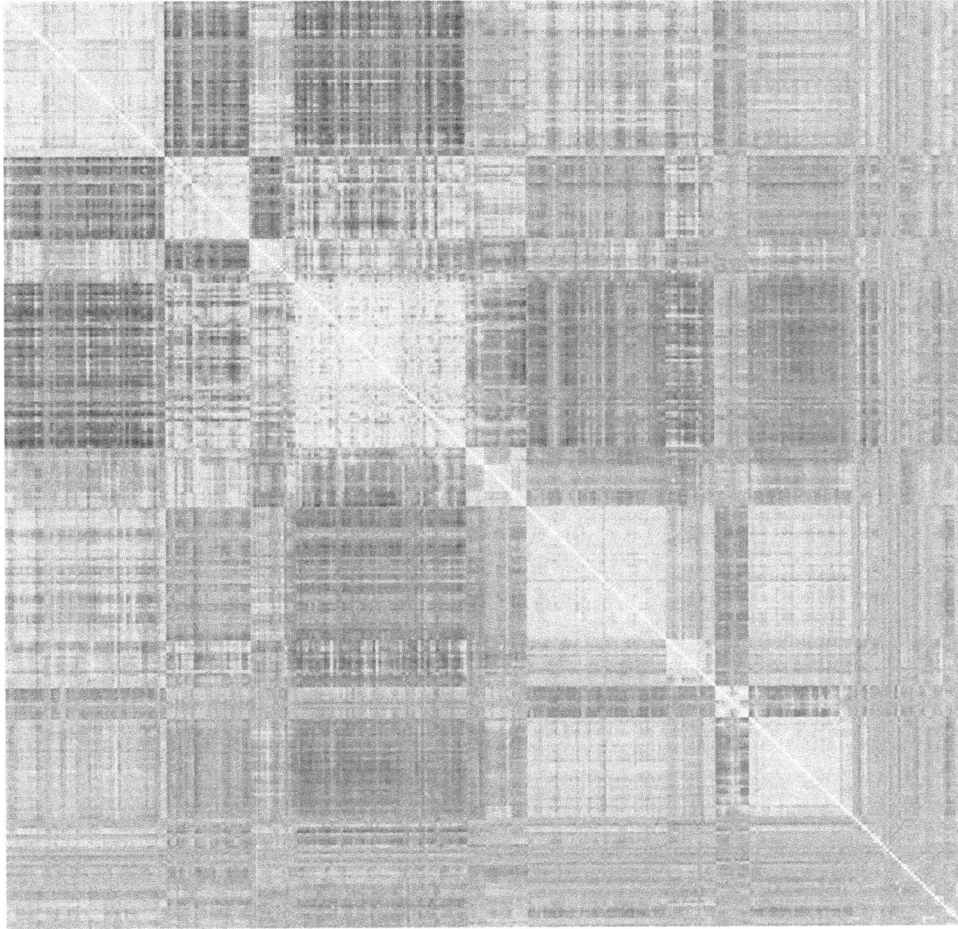

FIGURE 16.3. A heatmap showing the correlation coefficient of brain activity between 316 regions in the left hemisphere of a single individual. Cells in light gray reflect strong positive correlation, whereas cells in dark gray reflect strong negative correlation. The large blocks of positive correlation along the diagonal of the matrix correspond to the major connected networks in the brain.

$$d(x, y) = \sqrt{(x_1 - x_2)^2 + (y_1 - y_2)^2}$$

Plugging in the values from our example data, we have

$$d(x, y) = \sqrt{(1 - 4)^2 + (2 - 3)^2} = 3.16$$

If the formula for Euclidean distance in two dimensions seems slightly familiar, this is because it is identical to the *Pythagorean theorem* that most of us learned in geometry class, which computes the length of the hypotenuse of a right triangle based on the lengths of

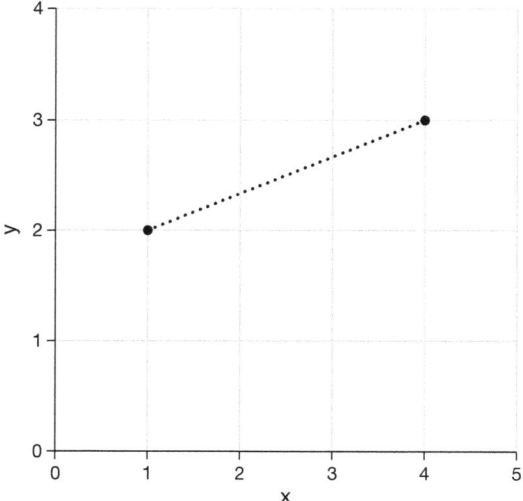

FIGURE 16.4. A depiction of the Euclidean
distance between two points, (1,2) and (4,3).
The two points differ by 3 along the *x*-axis and by
1 along the *y*-axis.

the two sides. In this case, the length of the sides of the triangle correspond to the distance between the points along each of the two dimensions. While this example is in two dimensions, the same idea extends to arbitrary numbers of dimensions.

One important feature of the Euclidean distance is that it is sensitive to the overall mean and variability of the data. In this sense, it is unlike the correlation coefficient, which measures the linear relationship between variables in a way that is insensitive to the overall mean or variability of each variable. For this reason, it is common to *scale* the data prior to computing the Euclidean distance, which is equivalent to converting each variable into its Z-scored version.

K-means Clustering

One commonly used method for clustering data is *K-means clustering*. This technique identifies a set of cluster centers and then assigns each data point to the cluster whose center is the closest (i.e., has the lowest Euclidean distance) from the data point. As an example, let's take the latitude and longitude of a number of countries around the world as our data points and see whether K-means clustering can effectively identify the continents of the world.

Most statistical software packages have a built-in function for performing K-means clustering using a single command, but it's useful to understand how it works step-by-step. We must first decide on a specific value for *K*, the number of clusters to be found in the data. It's important to point out that there is no unique, "correct" value for the number of clusters; there are various techniques that one can use to try to determine which solution

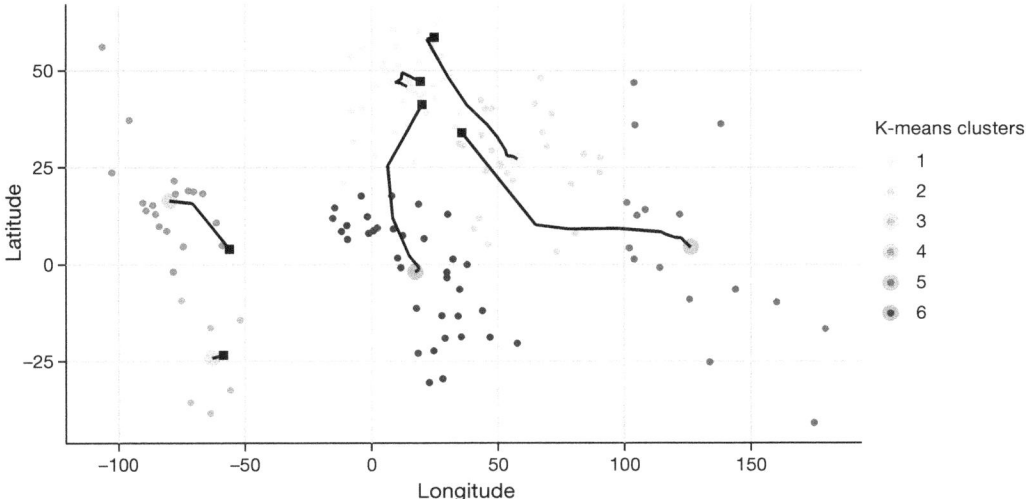

FIGURE 16.5. A two-dimensional depiction of clustering on the latitude and longitude of countries across the world. The square symbols show the starting centroids for each cluster, and the lines show the movement of the centroid for that cluster across the iterations of the algorithm, which are presented with larger points.

is "best," but they can often give different answers, as they incorporate different assumptions or trade-offs. Nonetheless, clustering techniques such as K-means are an important tool for understanding the structure of data, especially as they become high-dimensional.

Having selected the number of clusters (K) that we wish to find, we must come up with K locations that will be our starting guesses for the centers of our clusters (since we don't initially know where the centers are). One simple way to start is to choose K of the actual data points at random and use those as our starting points, which are referred to as *centroids*. We then compute the Euclidean distance of each data point to each of the centroids and assign each point to a cluster based on its closest centroid. Using these new cluster assignments, we recompute the centroid of each cluster by averaging the location of all the points assigned to that cluster. This process is then repeated until a stable solution is found; we refer to this as an *iterative* process, because it iterates until the answer doesn't change, or until some other kind of limit is reached, such as a maximum number of allowable iterations.

Applying K-means clustering to the latitude/longitude data (figure 16.5), we see that there is a reasonable match between the resulting clusters and the continents, though none of the continents is perfectly matched to any of the clusters. We can further examine this by plotting a table that compares the membership of each cluster to the actual continents for each country, as shown in table 16.1; this kind of table is often called a *confusion matrix*.

- Cluster 1 contains all European countries, as well as countries from northern Africa and Asia.
- Cluster 2 contains contains Asian countries as well as several African countries.

Table 16.1. Comparison of K-means Clustering Result to Actual Continents

Cluster	Africa	Asia	Europe	North America	Oceania	South America
1	5	1	36	0	0	0
2	3	24	0	0	0	0
3	0	0	0	0	0	7
4	0	0	0	15	0	4
5	0	10	0	0	6	0
6	35	0	0	0	0	0

- Cluster 3 contains countries from the southern part of South America.
- Cluster 4 contains all of the North American countries as well as northern South American countries.
- Cluster 5 contains Oceania as well as several Asian countries.
- Cluster 6 contains all of the remaining African countries.

Although in this example we know the actual clusters (i.e., the continents of the world), in general we don't actually know the ground truth for unsupervised learning problems, so we simply have to trust that the clustering method has found useful structure in the data. However, one important point about K-means clustering, and iterative procedures in general, is that they are not guaranteed to give the same answer each time they are run. The use of random numbers to determine the starting points means that the starting points can differ each time; and, depending on the data, this can sometimes lead to different solutions being found. For this example, K-means clustering will sometimes find a single cluster encompassing both North and South America and sometimes find two clusters (as it did for the specific choice of random seed used here). Whenever using a method that involves an iterative solution, it is important to rerun the method a number of times using different random seeds, to make sure that the answers don't diverge too greatly between runs. If they do, then one should avoid making strong conclusions based on the unstable results. In fact, it's probably a good idea to avoid strong conclusions on the basis of clustering results more generally; they are primarily useful for getting intuition about the structure that might be present in a dataset.

We can apply K-means clustering to the self-control variables in order to determine which variables are most closely related to one another. For $K = 2$, the K-means algorithm consistently picks out one cluster containing the SSRT variables and one containing the impulsivity variables. With higher values of K, the results are less consistent; for example, with $K = 3$ the algorithm sometimes identifies a third cluster containing only the UPPS sensation-seeking variable, whereas in other cases it splits the SSRT variables into two separate clusters (as seen in figure 16.6). The stability of the clusters with $K = 2$ suggests that this is probably the most robust clustering for these data, but these results also highlight the importance of running the algorithm multiple times to determine whether any particular clustering result is stable.

SSRT_motor	SSRT_stim	SSRT_high	SSRT_low	UPPS_senseek	UPPS_posurg	UPPS_negurg	UPPS_pers	UPPS_premed	
1	1	1	1	3	2	2	2	2	3
1	1	1	1	3	2	2	2	2	1
1	1	1	1	3	2	2	2	2	4
1	1	1	1	3	2	2	2	2	5
1	1	1	1	3	2	2	2	2	6
1	1	1	1	3	2	2	2	2	7
1	1	1	1	3	2	2	2	2	8
1	1	1	1	3	2	2	2	2	9
1	1	3	3	2	2	2	2	2	10
1	1	3	3	2	2	2	2	2	2

FIGURE 16.6. A visualization of the clustering results from 10 runs of the K-means clustering algorithm with $K = 3$. Each row in the figure represents a different run of the clustering algorithm (with different random starting points), and variables sharing the same shade of gray are members of the same cluster. The numbers to the right represent the run numbers for each result.

Hierarchical Clustering

Another useful method for examining the structure of a multivariate dataset is known as *hierarchical clustering*. This technique also uses the distances between data points to determine clusters, but it additionally provides a way to visualize the relationships between data points in terms of a treelike structure known as a *dendrogram*.

The most commonly used hierarchical clustering procedure is known as *agglomerative clustering*. This procedure starts by treating each data point as its own cluster, and then progressively creates new clusters by combining the two clusters with the least distance between them. It continues to do this until there is only a single cluster. This requires computing the distance between clusters, and there are numerous ways to do this; in this example, we use the *average linkage* method, which simply takes the average of all the distances between each data point in each of two clusters. The example examines the relationship between the self-control variables that were described above.

Figure 16.7 shows the dendrogram generated from the self-regulation dataset. Here we see that there is structure in the relationships between variables that can be understood at various levels by "cutting" the tree at different distance values to create different numbers of clusters: if we cut the tree at 25, we get two clusters; if we cut it at 20, we get three clusters, and at 19 we get four clusters. Interestingly, the solution found by the hierarchical clustering analysis of the self-control data is identical to the solution found in the majority of the K-means clustering runs, which is comforting.

Our interpretation of this analysis is that there is a high degree of similarity within each of the variable sets (SSRT and UPPS) compared to between the sets. Within the UPPS

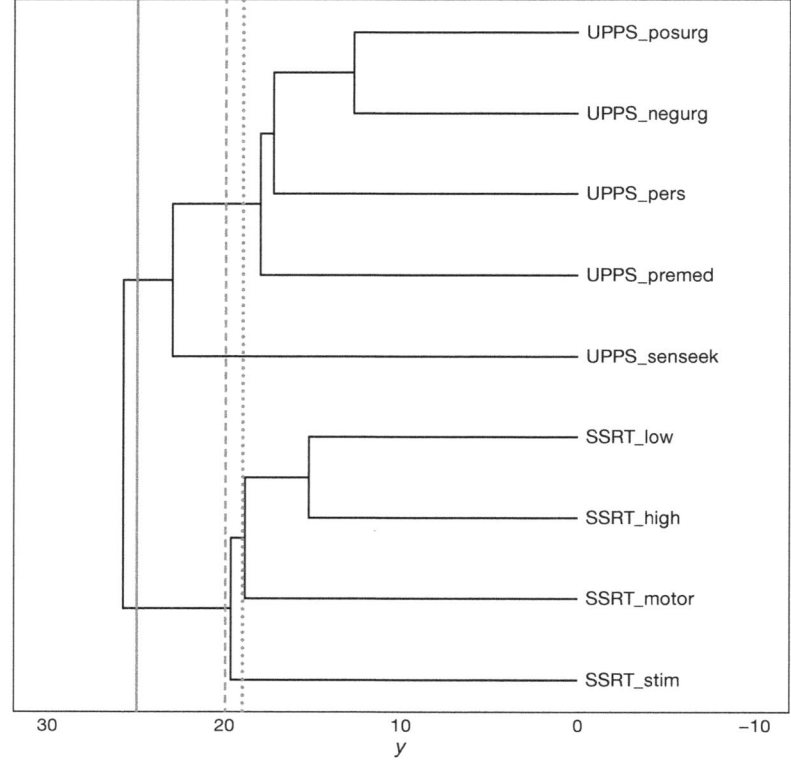

FIGURE 16.7. A dendrogram depicting the relative similarity of the nine self-control variables. The three vertical lines represent three different cutoffs, resulting in either two (solid line), three (dashed line), or four (dotted line) clusters.

variables, it seems that the sensation-seeking variable stands separately from the others, which are much more similar to one another. Within the SSRT variables, it seems that the stimulus-selective SSRT variable is distinct from the other three, which are more similar. These are the kinds of conclusions that can be drawn from a clustering analysis. It is again important to point out that there is no single "right" number of clusters; different methods rely upon different assumptions or heuristics and can give different results and interpretations. In general, it's good to present the data clustered at several different levels and to make sure that this doesn't drastically change one's scientific conclusions.

Dimensionality Reduction

It is often the case with multivariate data that many of the variables will be highly similar to one another, such that they are largely measuring the same thing. One way to think of this is that, while the data have a particular number of variables (which we call their *dimensionality*), in reality there are not as many underlying sources of information as there are variables. The idea behind *dimensionality reduction* is to reduce the number of variables in order to create composite variables that reflect underlying signals in the data.

Principal Component Analysis

The idea behind principal component analysis (commonly abbreviated as PCA) is to find a lower-dimensional description of a set of variables that accounts for the maximum possible amount of information in the full dataset. A deep understanding of principal component analysis requires an understanding of linear algebra, which is beyond the scope of this book; see the Suggested Reading section at the end of this chapter for helpful guides to this topic. In this section, I outline the concept in a nonmathematical way and hopefully whet your appetite to learn more.

We start with a simple example with just two variables in order to give an intuition for how PCA works. First, we generate some synthetic data for variables x and y, with a correlation of 0.7 between the two variables. The goal of principal component analysis is to find a linear combination of the observed variables in the dataset that will explain the maximum amount of variance; the idea here is that the variance in the data is a combination of signal and noise, and we want to find the strongest common signal between the variables. The first principal component is the combination that explains the maximum variance. The second component is the one that explains the maximum remaining variance, while also being uncorrelated with the first component. With more variables, we can continue this process to obtain as many components as there are variables (assuming that there are more observations than there are variables), though in practice we usually hope to find a small number of components that can explain a large portion of the variance.

In the case of our two-dimensional example, we can compute the principal components and plot them over the data (figure 16.8). What we see is that the first principal component (shown in light gray green) follows the direction of the greatest variance. This line is similar, though not identical, to the linear regression line; while the linear regression solution minimizes the distance between each data point and the regression line at the same x value (i.e., the vertical distance), the principal component minimizes the Euclidean distance between the data points and the line representing the component (i.e., the distance perpendicular to the component). The second component points in a direction that is perpendicular to the first component (which is the geometric equivalent of being uncorrelated).

It's common to use PCA to reduce the dimensionality of a more complex dataset. For example, let's say that we would like to know whether performance on all four of the stop-signal task variables in the self-control dataset is related to the five impulsivity survey variables. We can perform PCA on each of those datasets separately and examine how much of the variance in the data is accounted for by the first principal component, which will serve as our summary of the data.

We see in figure 16.9 that, for the stop-signal variables, the first principal component accounts for about 60% of the variance in the data, whereas for the UPPS it accounts for about 55% of the variance. This suggests that the first component is a reasonable summary for each set of variables. We can then compute the correlation between the scores obtained using the first principal component from each set of variables, in order to ask whether

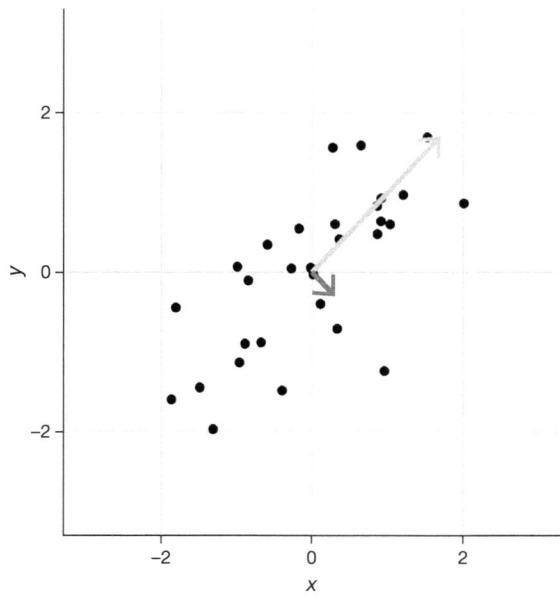

FIGURE 16.8. A plot of synthetic data, with the first principal component plotted in light gray and the second in dark gray.

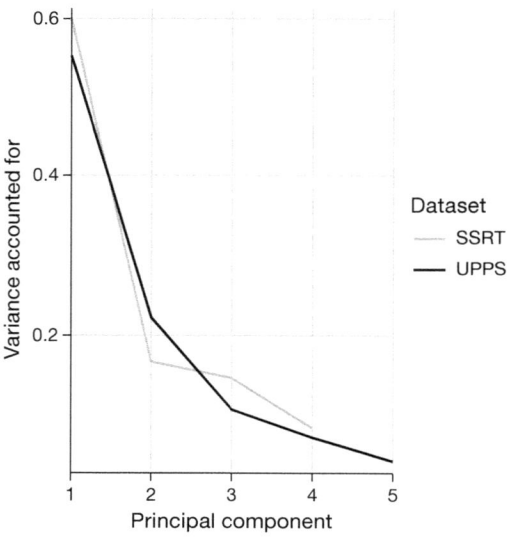

FIGURE 16.9. A plot of the variance accounted for with PCA applied separately to the response inhibition and impulsivity variables from the self-control dataset.

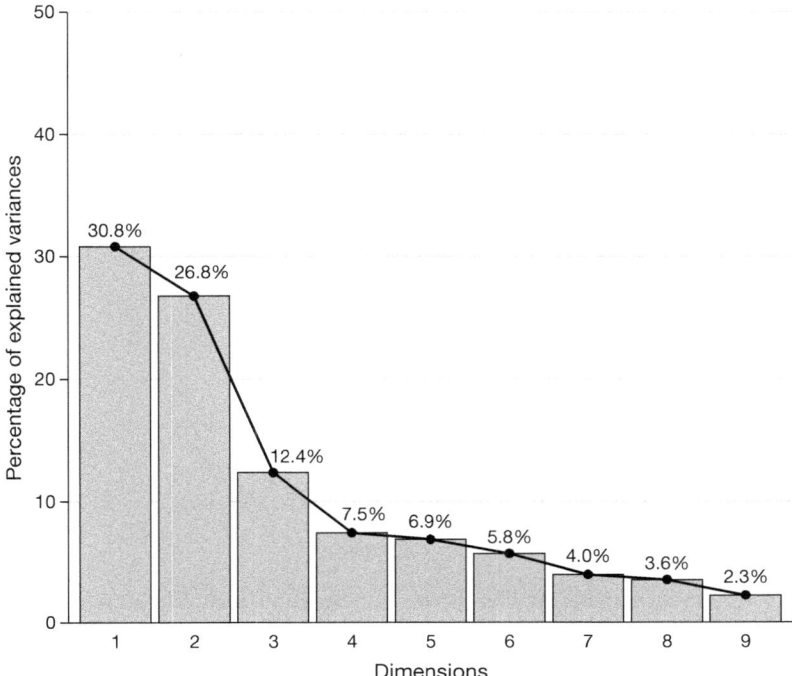

FIGURE 16.10. Plot of variance accounted for by PCA components computed on the full set of self-control variables.

there is a relationship between the two sets of variables. The correlation of −0.014 between the two summary variables suggests that there is no overall relationship between response inhibition and impulsivity in this dataset. The fact that the confidence interval for the correlation is fairly narrow also suggests that the data are consistent with a true correlation that is close to zero.

```
##
##   Pearson's product-moment correlation
##
## data:   SSRT and UPPS
## t = -0.3, df = 327, p-value = 0.8
## alternative hypothesis: true correlation is not equal to 0
## 95 percent confidence interval:
##   -0.123  0.093
## sample estimates:
##     cor
## -0.015
```

We could also perform PCA on all of these variables at once. Looking at the plot of variance accounted for (also known as a *scree plot*) in figure 16.10, we can see that the

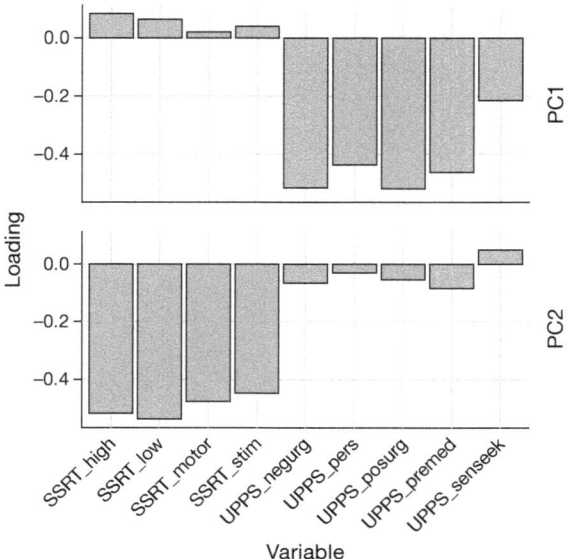

FIGURE 16.11. Plot of variable loadings in the PCA solution, including all self-control variables. Each variable is shown in terms of its loadings on each of the two components (noted on the right).

first two components account for a substantial amount of the variance in the data. We can then look at the loadings of each of the individual variables on these two components to understand how each specific variable is associated with the different components.

Doing this for the self-control dataset (figure 16.11), we see that the first component (in the first row of the figure) has nonzero loadings for most of the UPPS variables and near-zero loadings for each of the SSRT variables, whereas the opposite is true of the second principal component, which loads primarily on the SSRT variables. This tells us that the first principal component captured mostly variance related to the impulsivity measures, while the second component captured mostly variance related to the response inhibition measures. You might notice that the loadings are actually negative for most of these variables; the sign of the loadings is arbitrary, so we should make sure to look at large positive and negative loadings.

Factor Analysis

While principal component analysis can be useful for reducing a dataset to a smaller number of composite variables, the standard method for PCA has some limitations. First, and most important, it ensures that the components are uncorrelated. While this may sometimes be useful, there are often cases where we want to extract dimensions that may be correlated with one another. Second, PCA doesn't account for measurement error in the variables that are being analyzed, which can lead to difficulty in interpreting the resulting loadings on the components. Although there are modifications of PCA that can address these issues, it is more common in some fields to use a technique called *exploratory factor analysis* (or EFA) to reduce the dimensionality of a dataset.[1]

The idea behind EFA is that each observed variable is created through a combination of contributions from a set of *latent variables*—that is, variables that cannot be observed directly—along with some amount of measurement error for each variable. For this reason, EFA models are often referred to as belonging to a class of statistical models known as *latent variable models*.

For example, let's say that we want to understand how measures of several different variables relate to the underlying factors that give rise to those measurements. We will first generate a synthetic dataset to show how this might work. This dataset contains a set of individuals for whom we will pretend that we know the values of several latent psychological variables: impulsivity, working memory capacity, and fluid reasoning. We will assume that working memory capacity and fluid reasoning are correlated with one another but that neither is correlated with impulsivity. From these latent variables, we will then generate a set of eight observed variables for each individual, which are simply linear combinations of the latent variables, along with random noise added to simulate measurement error.

We can further examine the data by displaying a heatmap of the correlation matrix relating all of these variables (figure 16.12). We see from this display that there are three clusters of variables corresponding to our three latent variables, which is as it should be.

We can think of EFA as estimating the parameters of a set of linear models all at once, where each model relates one of the observed variables to all of the latent variables. For our example, the equations would look like those shown below. In these equations, the β characters have two subscripts—one that refers to the task and the other that refers to the latent variable—and a variable ϵ that refers to the error. Here we assume that everything has a mean of zero, so that we don't need to include an extra intercept term for each equation.

$$
\begin{aligned}
nback &= beta_{[1,1]} * WM + \beta_{[1,2]} * FR + \beta_{[1,3]} * IMP + \epsilon \\
dspan &= beta_{[2,1]} * WM + \beta_{[2,2]} * FR + \beta_{[2,3]} * IMP + \epsilon \\
ospan &= beta_{[3,1]} * WM + \beta_{[3,2]} * FR + \beta_{[3,3]} * IMP + \epsilon \\
ravens &= beta_{[4,1]} * WM + \beta_{[4,2]} * FR + \beta_{[4,3]} * IMP + \epsilon \\
crt &= beta_{[5,1]} * WM + \beta_{[5,2]} * FR + \beta_{[5,3]} * IMP + \epsilon \\
UPPS &= beta_{[6,1]} * WM + \beta_{[6,2]} * FR + \beta_{[6,3]} * IMP + \epsilon \\
BIS11 &= beta_{[7,1]} * WM + \beta_{[7,2]} * FR + \beta_{[7,3]} * IMP + \epsilon \\
dickman &= beta_{[8,1]} * WM + \beta_{[8,2]} * FR + \beta_{[8,3]} * IMP + \epsilon
\end{aligned}
$$

In effect, what we want to do using EFA is to estimate the *matrix* of coefficients (betas) that map the latent variables into the observed variables. For the data that we are

---

1. There is another application of factor analysis known as *confirmatory factor analysis* (or CFA) that we do not discuss here. In practice, its application can be problematic, and recent work has started to move toward modifications of EFA that can answer the questions often addressed using CFA (Marsh et al. 2014).

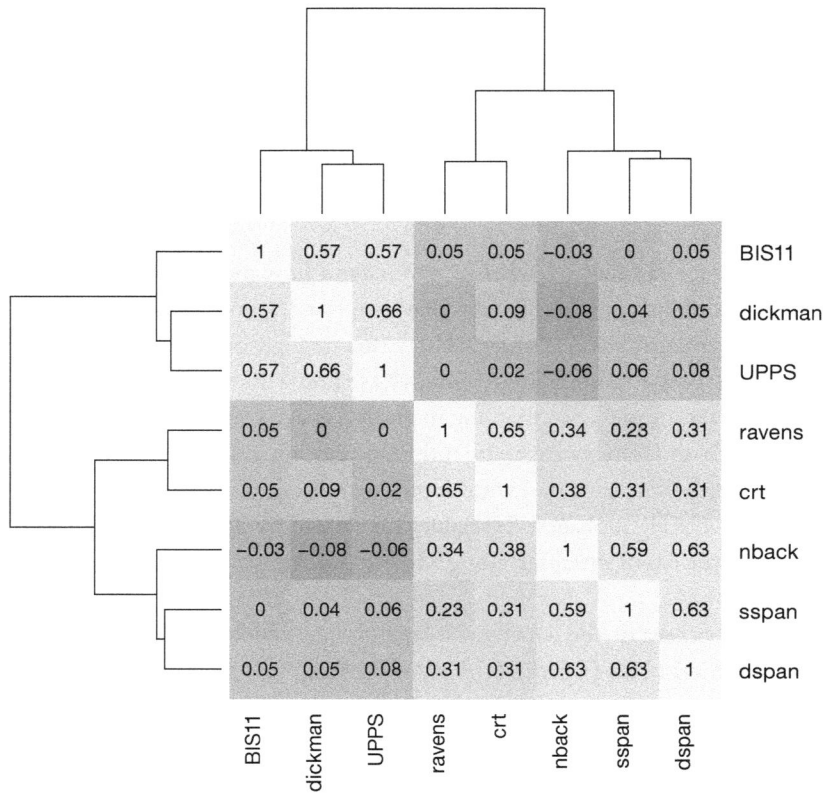

FIGURE 16.12. A heatmap showing the correlations between the variables generated from the three underlying latent variables. The variables have been ordered based on their cluster, with dendrograms (shown at the top and to the left) describing the clustering of the variables.

generating, we know that most of the betas in this matrix are zero because we created them that way; for each task, only one of the weights is set to one, which means that each task is a noisy measurement of a single latent variable.

We can apply EFA to our synthetic dataset to estimate these parameters. We won't go into the details of how EFA is actually performed, other than to mention one important point. Most of the previous analyses in this book have relied upon methods that try to minimize the difference between the observed data values and the values predicted by the model. The methods that are used to estimate the parameters for EFA instead attempt to minimize the difference between the *covariances* of the observed variables and the covariances implied by the model parameters. For this reason, these methods are often referred to as *covariance structure models*.

Let's apply an exploratory factor analysis to our synthetic data. As with clustering methods, we need to first determine how many latent factors we want to include in our model. In this case, we know there are three factors, so let's start with that; later on we will examine ways to estimate the number of factors directly from the data. Here is the output from our statistical software for this model:

```
##
## Factor analysis with Call: fa(r = observed_df, nfactors = 3)
##
## RMSEA index =   0.026
## BIC =   -29
##  With factor correlations of
##      MR1   MR2   MR3
## MR1 1.00 0.03 0.47
## MR2 0.03 1.00 0.03
## MR3 0.47 0.03 1.00
```

One question that we would like to ask is how well our model actually fits the data. There is no single way to answer this question; rather, researchers have developed a number of different methods that provide some insight into how well the model fits the data. For example, one commonly used criterion is based on the *root mean squared error of approximation* (RMSEA) statistic, which quantifies how far the predicted covariances are from the actual covariances; a value of RMSEA less than 0.08 is often considered to reflect an adequately fitting model. In the example here, the RMSEA value is 0.026, which suggests a model that fits quite well.

We can also examine the parameter estimates in order to see whether the model has appropriately identified the structure in the data. It's common to plot this as a graph, with arrows from the latent variables (represented as ellipses) pointing to the observed variables (represented as rectangles), where an arrow represents a substantial loading of the observed variable on the latent variable; this kind of graph is often referred to as a *path diagram*, since it reflects the paths relating the variables. This is shown in figure 16.13. In this case, the EFA procedure correctly identified the structure present in the data, both in terms of which observed variables are related to each of the latent variables and in terms of the correlations between latent variables.

Determining the Number of Factors

One of the main challenges in applying EFA is to determine the number of factors. A common way to do this is to examine the fit of the model while varying the number of factors, and then selecting the model that gives the best fit. This is not foolproof, and there are multiple ways to quantify the fit of the model that can sometimes give different answers.

One might think that we could simply look at how well the model fits and pick the number of factors with the best fit, but this won't work because a more complex model will always fit the data better (as we saw in our earlier discussion of overfitting). For this reason, we need to use a metric of model fit that includes a penalty for the number of parameters in the model. For the purposes of this example, we select one of the common

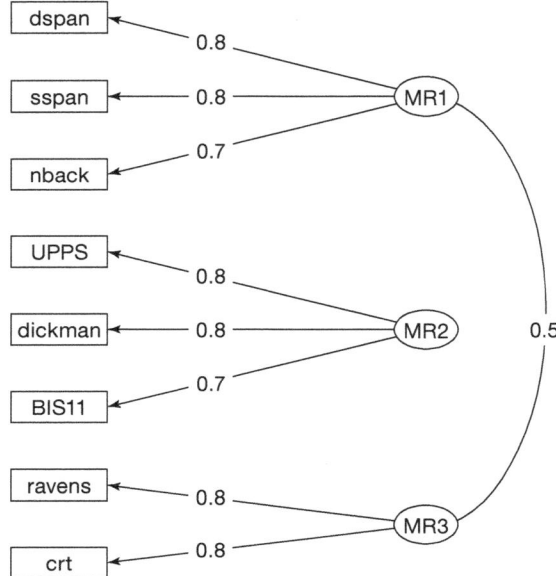

FIGURE 16.13. Path diagram for the exploratory factor analysis model.

methods for quantifying model fit, which is known as the *sample-size adjusted Bayesian information criterion* (or SABIC). This measure quantifies how well the model fits the data (by asking how likely the data are, given the model), while also taking into account the number of parameters in the model (which in this case is related to the number of factors) as well as the sample size. While the absolute value of SABIC is not interpretable, when using the same data and the same kind of model, we can compare models using SABIC to determine which is most appropriate for the data. One important thing to know about SABIC and other measures like it (known as *information criteria*) is that lower values represent a better fit of the model, so in this case we want to find the number of factors with the lowest SABIC. In figure 16.14, we see that the model with the lowest SABIC has three factors, which shows that this approach was able to accurately determine the number of factors used to generate the data.

Now let's see what happens when we apply this model to real data from the self-control dataset, which contains measurements of all eight variables that were simulated in the example above. The model with three factors also has the lowest SABIC for these real data. Plotting the path diagram (figure 16.15), we see that the real data demonstrate a factor structure that is very similar to what we saw with the simulated data. This should not be surprising, since the simulated data were generated based on knowledge of these different tasks; but it's comforting to know that human behavior is systematic enough that we can reliably identify these kinds of relationships. The main difference is that the correlation between the working memory factor (MR3) and the fluid reasoning factor (MR1) is even higher than it was in the simulated data. This result is scientifically useful because it shows

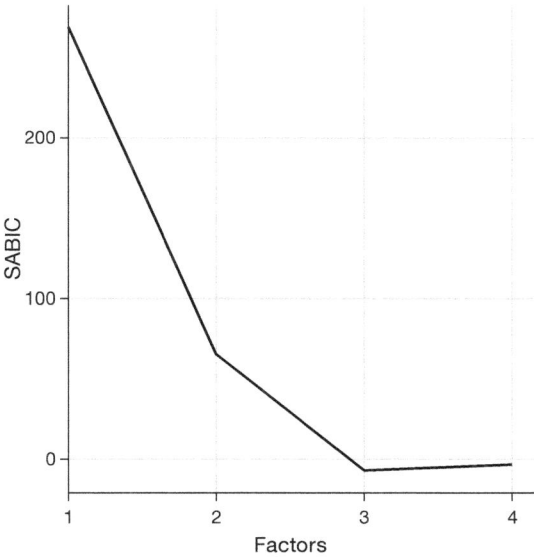

FIGURE 16.14. Plot of SABIC for varying numbers of factors.

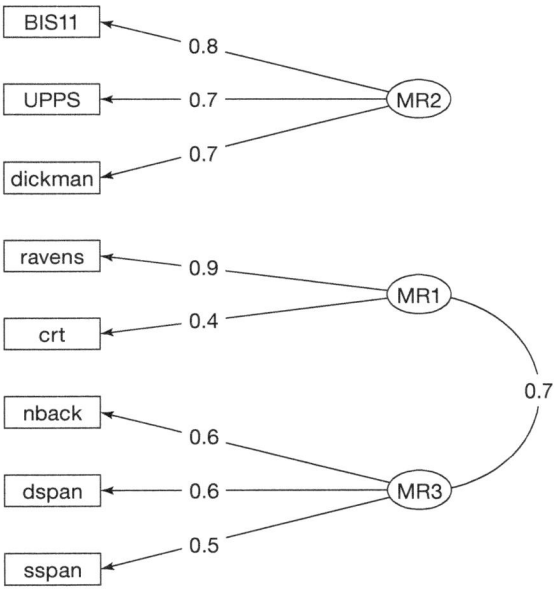

FIGURE 16.15. Path diagram for the three-factor model on the self-control dataset.

us that, while working memory and fluid reasoning are closely related, there is utility in separately modeling them.

Suggested Reading

- *The Geometry of Multivariate Statistics*, by Thomas Wickens. This classic book by the late Tom Wickens shows how to think about multivariate statistics from a geometric standpoint.
- *No Bullshit Guide to Linear Algebra*, by Ivan Savov. This book lives up to its name in providing a very clear yet detailed overview of linear algebra, including all of the concepts needed to understand the methods discussed in this chapter.
- "Clustering: Science or Art?" by Ulrike von Luxburg, Robert Williamson, and Isabelle Guyon. This paper (freely available online) lays out exactly why there is generally no "right" answer when it comes to clustering of data.

Problems

1. Describe the rationale for using a scatterplot of matrices or a heatmap to visualize multivariate data. When would one be more appropriate than the other?
2. What is an iterative analysis method, and why might it be important to run it multiple times and compare the results? Which methods discussed in this chapter were described as iterative?
3. Describe the K-means clustering algorithm.
4. How can we determine the correct number of clusters for a clustering problem?
    - By determining which clustering solution accounts for the most variance in the data.
    - By visually inspecting the solutions and finding the one that looks most pleasing.
    - By computing the similarity within the clusters and finding the number that maximizes this similarity.
    - There is no single way to determine the correct number of clusters.
5. Which of the following are true of the results from principal component analysis? Choose all that apply.
    - The first principal component accounts for more variance than the second component.
    - The principal components are uncorrelated with one another.
    - The principal component is identical to the regression line.
    - The sign of the component values is arbitrary.
6. Describe one important way that exploratory factor analysis differs from principal component analysis.
7. Describe a method that can be used to determine the number of factors in an exploratory factor analysis.

17

Practical Statistical Modeling

In this chapter, you will see several examples of statistical analyses applied to real data. While the foregoing chapters have provided you with the basis for understanding how to analyze data, things often become messy when working with real data. In this chapter, we lay out an overall strategy for performing a statistical analysis and provide several examples. The examples come from a broad range of domains, to help demonstrate the many kinds of questions that can be addressed using statistical modeling. We also introduce some more advanced analysis methods and concepts, which are necessary to properly analyze some of the datasets, including logistic regression and mixed-effects models, and we delve into greater detail about how to perform diagnostics on the results from statistical analyses, along with ways to address commonly encountered issues.

Learning Objectives

Having read this chapter, you should be able to

- Determine the appropriate statistical model to test a particular scientific hypothesis.
- Identify outliers and describe why they could be problematic for an analysis.
- Determine whether a mixed-effects model is necessary in a particular situation.
- Check the assumptions of the statistical model to ensure that they are not violated.

The Process of Statistical Modeling

There is a set of steps that we generally go through when we want to use our statistical model to test a scientific hypothesis:

1. Specify the question of interest.
2. Identify or collect the appropriate data.
3. Prepare and visualize the data.
4. Determine the appropriate model.
5. Fit the model to the data.

6. Perform diagnostics on the model to check assumptions.
7. Test hypotheses and quantify effect size.

In each of the following examples, we will follow these steps to perform the analysis. Note that sometimes one might have to circle back and perform some steps multiple times; for example, we usually start by fitting a relatively simple initial model and then find that some of the important assumptions of the model have been violated, in which case we may need to modify the model and/or transform the data and then refit the model.

Data Wrangling

For each of the analyses described below, the raw data require some degree of manipulation (often called "wrangling") to get them into a form where they are ready for data analysis. The form of wrangling differs depending on the particular language and software packages to be used, but it often involves steps such as changing the names of variables, combining data across multiple files or data sources, and modifying the format of variables. In the examples below, we do not delve into the details of this process, but interested readers can see the entire preparation process in the code available via https://press.princeton .edu/isbn/9780691218441, under Resources.

Gaining facility with data wrangling is an important aspect of real-world statistical practice, and the best way to do it is using a programming language such as R or Python. It's often easy to do simple data wrangling using a spreadsheet program like Microsoft Excel, but the results can be impossible to reproduce or check, often leading to errors. A striking example occurred in a well-known economics publication (Reinhart and Rogoff 2010), which had argued that national debt was associated with decreased economic growth; this paper was instrumental in the austerity programs that led to substantial misery across the world in the early 2010s. It turned out that their conclusion was flawed due to an error that was made while working with the data in an Excel spreadsheet.[1]

Specifying the Statistical Model

Perhaps the most challenging aspect of statistical analysis is determining the appropriate model and then specifying the variables that should appear in the model. Here we outline some of the important factors in determining and specifying the model.

How is the Dependent Variable Distributed?

The distribution of the dependent variable is very important in determining the statistical model to be used, because it has a major impact on the distribution of the residuals. Most statistical methods require that we make assumptions about the distribution of residuals

---

1. https://theconversation.com/the-reinhart-rogoff-error-or-how-not-to-excel-at-economics-13646.

in order to perform accurate statistical inference, and it is thus critical to ensure that the required assumptions are met.

When the data are continuously distributed, we generally take the linear regression model as our starting point. For this model, the residuals from the model must be (at least roughly) normally distributed in order for the p-values from the model to be trusted. Importantly, this assumption is particularly critical when the sample sizes are relatively low; once the sample size reaches into the hundreds or thousands, the central limit theorem ensures that the sampling distributions of our statistics should be normal, even if the residuals are not. For all of the analyses below that use linear regression models, we will check the residuals using quantile-quantile (Q-Q) plots to look for obvious deviations from normality. There is no strict cutoff at which deviations from normality become problematic, but if deviations are obvious then they should be addressed, especially if the sample size is relatively small.

In each of the analyses below, we will look at the distribution of the data before specifying the model. Even though our assumption of normality applies to the residuals from the model rather than the raw data, it is often true that deviations from normality will arise from nonnormality in the data. For instance, in example 1, we will encounter a dataset where the dependent variable is binary (that is, having values of zero or one). If we fit a linear regression model to these data, it will be basically impossible to obtain normally distributed residuals. This is also likely to be true if the data are very strongly skewed or have a long-tailed distribution.

When the data have a structure that causes the residuals to be nonnormal, we have two options. First, we can apply a mathematical transformation to the data to change the shape of the distribution. When the data are skewed, a transform such as a logarithm or square root is often sufficient to make the distribution roughly normal, as we will see in example 4. In other cases, we might decide to fundamentally transform the data so that a different type of model is appropriate. In example 1, we transform the data from a highly skewed variable with a large number of zeros into a binary variable. Second, we can use a different type of statistical model. In example 1, we introduce the *logistic regression* model, which is an extension of linear regression that is appropriate for binary data. There are also much more sophisticated models that can be used with other types of data, but these are outside of the scope of this book and generally require consultation with an expert statistician.

Finally, it is important to note that nonnormality of residuals can often occur when the statistical model does not include important variables of interest, which we refer to as *misspecification* of the model. For example, imagine that we were analyzing a dataset involving the comparison of scores on a creative writing test between two different college programs. Unbeknownst to us, the data include scores from two groups of students: English majors and math majors. If the statistical model does not include a variable accounting for the differences between the majors, this could result in nonnormality of the residuals if the writing test scores of the two groups are very different. Thus, when nonnormal residuals are encountered, one should always consider whether the model is properly specified and whether there are any important variables that might be missing.

Beyond normality, we also generally need to assume that the residuals are independent and have the same variance across observations; this assumption is often referred to as *independent and identically distributed*, or IID. Hidden structure in the data can often result in violations of this assumption, which can cause biased effect estimates as well as flawed inferences. For example, imagine that we are analyzing the performance of students in an elementary school on a reading test. If some teachers are better than others, but we do not include this in the model, then the students with better teachers will generally perform better (and thus have positive residuals), while those with worse teachers will have negative residuals due to lower performance. This can similarly occur when the same individual is tested repeatedly on the same measurement, where observations are likely be more similar within versus between individuals; this is the same issue that we saw in the paired *t* test in chapter 15. The presence of this kind of structure in the data is often referred to as *clustered data*, and it is a very common problem that will be seen in examples 2, 3, and 4 below. There are a number of different techniques that are used to address this problem across different research domains. We introduce a particular method known as *mixed-effects modeling* that is commonly used in a number of fields to address clustered data.

Outliers and Influential Observations

We have already seen the potential impact of outliers in chapter 13, where we discussed them in the context of correlations, and the same issue can arise in nearly any analysis of continuous outcomes. An outlier is a data point that lies far outside the overall distribution. There is no single criterion to distinguish outliers from regular data points; one commonly used definition comes from John Tukey, which is enshrined in the box plot that you have already encountered. Tukey's criterion is based on the *interquartile range* (IQR), which is the distance between the twenty-fifth percentile (or first quartile) and the seventy-fifth percentile (or third quartile) of the data (figure 17.1). An outlier is defined as any point that falls more than 1.5 IQRs more extreme than the first or third quartile (which is where the "whiskers" of the box plot end); these are shown using individual points in the box plot.

In the context of regression, we are particularly interested in the influence of any particular data point on the estimated parameters of the model. One way to conceptualize this is in terms of how much the regression parameters would change if each data point were omitted. A commonly used statistic, known as *Cook's distance* (often referred to as *Cook's D*), can be used to quantify this, and is available from most regression software. In figure 17.1, you can see that Cook's D is larger for data points that fall farther from the regression line and are toward the extremes of the *x* value (though the points don't necessarily have to be far from the regression line). The term *leverage* is often used to describe this, analogous to the physical phenomenon where the effect of a force depends on how far it is from the fulcrum (which in this case is the mean of the *x* variable). A rule of thumb

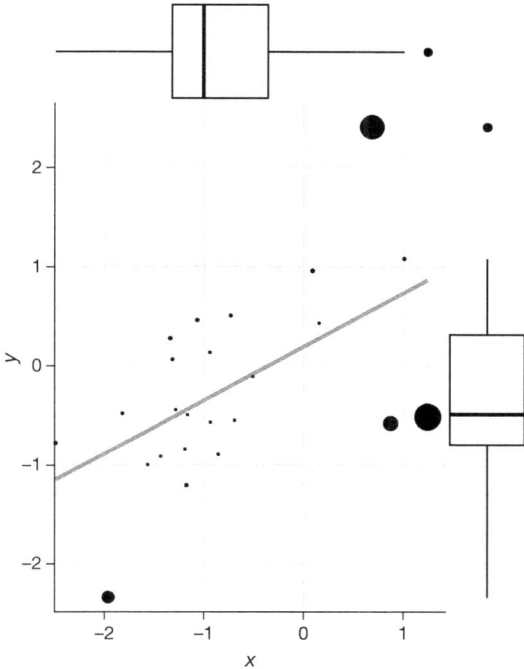

FIGURE 17.1. A scatterplot with box plots for each of the individual variables (known as *marginal distributions*). The relative influence (Cook's distance) of each data point is reflected by its size.

is that any observation with a Cook's D value greater than one should be investigated, but values with a smaller Cook's D can also have substantial influence on the regression solution. While highly influential points are often also extreme on the *y* value, this example shows how points can also be influential yet fall well within the middle of the *y* value distribution. The same is not true of the *x* variable: just as a person standing at the middle of a seesaw will have trouble moving it, data points in the center of the *x* distribution are unlikely to have high leverage.

When outliers are identified, it is important to consider how they might have arisen. In some cases, outliers can reflect a mistake in the measurement process; for example, in a study measuring height, a researcher might inadvertently enter the value in meters rather than centimeters, or might transpose the number 175 to result in 751. It is thus crucial to understand the variables that are being measured and ensure that all values fall within legitimate ranges. In one notable example, a published study (Lewandowsky, Gignac, and Oberauer 2013) analyzed the relationship between age and worldviews on a number of issues related to rejection of science, in which they reported that age was unrelated to these variables. They later issued a correction to to the article (Lewandowsky, Gignac, and Oberauer 2015), which revealed that "the dataset included two notable age outliers (reported ages 5 and 32757)." This is a case where appropriate visual examination of the data would certainly have revealed the presence of an invalid age value.

Outliers may also reflect the presence of more complex processes underlying the generation of the data. For example, it is common in psychology to measure response times in order to analyze psychological processes. In these studies, responses are often made in well under one second, but it is not uncommon to see occasional outlier response times lasting more than two seconds. In these cases, we generally think that these long responses reflect some kind of distraction or mind wandering, such that something different is happening on these trials compared to the rest of the trials. For many researchers, this justifies removing the offending outliers (as is common in the analysis of response times), since they may not be indicative of the same data-generating mechanisms. This decision must be based on a scientific understanding of the data and how they were generated, and must be made *before* an analysis of the data.

In yet other cases, outliers may simply reflect the underlying distribution of the data. For example, when a variable has a skewed long-tailed distribution, the most extreme values in the dataset are likely to show up as outliers in any analysis. However, removal of those extreme values would likely remove some of the most important data points and potentially bias any analysis. One solution in these cases is to transform the data; for example, a logarithmic transformation generally tends to pull in extreme observations that might otherwise have high leverage on the result.

What are the Independent Variables of Interest?

In general, if we know that a variable could potentially impact the value of the dependent variable, then we should include it in our model. One reason was just outlined above: failure to include a relevant variable could lead to nonrandom structure in the residuals, which can violate the assumptions of our model. A second reason has to do with our interpretation of the results, as a well-worn example can show us.

A researcher wishes to understand the factors that are associated with shark attacks, so she uses a conveniently available dataset to investigate this question. She builds a linear regression model that includes the local sales volume of ice cream as the independent variable and the number of shark attacks as the dependent variable. This analysis shows a highly significant effect of ice cream sales on the prevalence of shark attacks. How do we interpret this? One might do some mental gymnastics to come up with an explanation involving spilled ice cream in the water attracting sharks to the beach, but there is a much more obvious explanation: both shark attacks and ice cream sales go up when it's hot out and there are people at the beach. We say that the number of people at the beach is a *confounder* because it is a common cause of changes in both an independent variable (shark attacks) and a dependent variable (ice cream sales). Whenever there are confounder variables that are not included in our statistical model, then we cannot interpret the regression parameters as reflecting causal relationships between the independent and dependent variables.

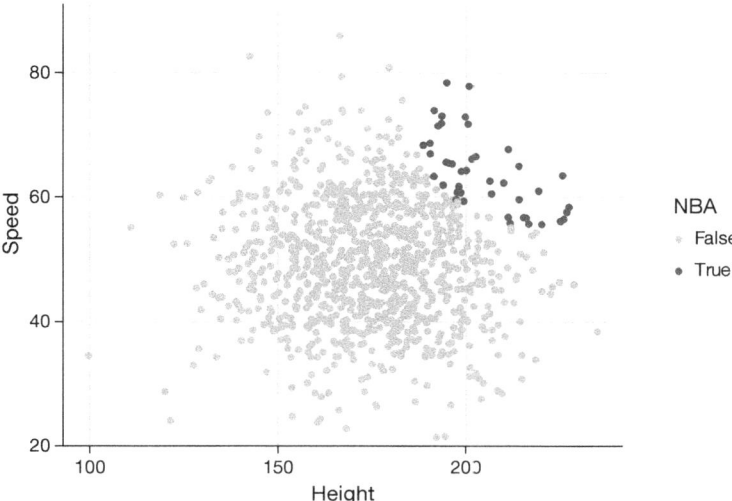

FIGURE 17.2. Scatterplot of simulated data for height versus speed of a group of basketball players. NBA players are selected based on a combination of speed and height and are shown in dark gray.

We should generally be wary of making any sort of *causal* interpretation of the results from a linear model, as outlined in chapter 13 in our discussion of causality and correlation. However, building our regression model *requires* that we think about possible causes of the dependent variables, for the following reason. Confounder variables are those that could potentially have a causal effect on both the independent and dependent variables (i.e., a *common cause*), and those are important to include in our model in order to ensure the proper interpretation of the regression parameters. But sometimes we might be tempted to include a variable in the model that is a *common effect* of the processes that are measured by both the dependent and other independent variables, and this can be highly problematic.

For example, say that we wish to assess whether height and speed are related to one another in basketball players.[2] Figure 17.2 shows a simulated dataset of height and speed, in which those two variables are unrelated, as we can see by fitting a linear regression model:

```
## 
## Call:
## lm(formula = speed ~ height, data = conf_df)
## 
## Coefficients:
##              Estimate Std. Error t value Pr(>|t|)
## (Intercept) 49.54246    2.86552   17.29   <2e-16 ***
## height       0.00216    0.01625    0.13     0.89
```

2. After https://observablehq.com/@herbps10/collider-bias.

```
## ---
## Signif. codes:  0 '***' 0.001 '**' 0.01 '*' 0.05 '.' 0.1 ' ' 1
##
## Residual standard error: 10 on 998 degrees of freedom
## Multiple R-squared:  1.77e-05,   Adjusted R-squared:  -0.000984
## F-statistic: 0.0176 on 1 and 998 DF,  p-value: 0.894
```

Now let's say that we wish to include an additional variable in the model that specifies which of the players are members of the NBA (shown in dark gray in the figure). We assume that NBA players fall into the top right corner of the plot; in order to make it into the NBA, one needs to be either very fast, very tall, or a good combination of both. If we include NBA player status in the model, now the effect of height on speed becomes significantly *negative*:

```
##
## Call:
## lm(formula = speed ~ height + NBA, data = conf_df)
##
## Coefficients:
##             Estimate Std. Error t value Pr(>|t|)
## (Intercept)  58.0959     2.8543   20.35   <2e-16 ***
## height       -0.0507     0.0163   -3.11   0.0019 **
## NBATRUE      16.0857     1.5775   10.20   <2e-16 ***
## ---
## Signif. codes:  0 '***' 0.001 '**' 0.01 '*' 0.05 '.' 0.1 ' ' 1
##
## Residual standard error: 9.7 on 997 degrees of freedom
## Multiple R-squared:  0.0945, Adjusted R-squared:  0.0926
## F-statistic:   52 on 2 and 997 DF,  p-value: <2e-16
```

The intuition here is that, while height and speed are unrelated in the general population, the selection of individuals for the NBA relies upon both height and speed, and NBA player status is thus a common effect of height and speed. This is often referred to as *collider bias*, and a graph node with multiple arrows pointing to it is referred to as a *collider*. If we were to draw a causal graph (see chapter 13), we would see both speed and height with arrows leading to NBA player status. This effect is also clear if we compute the same regression only on the NBA players, where we see a strong negative relationship between speed and height:

```
##
## Call:
## lm(formula = speed ~ height, data = conf_df_nba)
##
## Coefficients:
```

```
##              Estimate Std. Error t value Pr(>|t|)
## (Intercept) 131.6209    12.9493   10.16  6.9e-13 ***
## height       -0.3315     0.0632   -5.24  4.8e-06 ***
## ---
## Signif. codes:  0 '***' 0.001 '**' 0.01 '*' 0.05 '.' 0.1 ' ' 1
##
## Residual standard error: 4.8 on 42 degrees of freedom
## Multiple R-squared:  0.396,  Adjusted R-squared:  0.381
## F-statistic: 27.5 on 1 and 42 DF,  p-value: 4.79e-06
```

This form of collider bias is known as *selection bias*, and it occurs when inclusion in the dataset is related to one of the variables in the model. Selection bias can have important detrimental effects on results of statistical analyses. For example, Woodfine and Redelmeier 2015 showed how selection bias could cause one to conclude that wearing a helmet results in *worse* injuries in a motorcycle accident, when in reality helmet wearing is protective. In this case, the bias arose by only looking at individuals who were hospitalized.

This example highlights two issues to keep in mind when specifying and interpreting regression models. First, one should never include a variable in the model if it is a plausible common effect of both the dependent variable and one of the independent variables, as this can severely bias the estimates of the model. Second, we must be aware of the potential effects of selection bias, which requires understanding how our samples are collected and what biases might have been in play.

Example 1: Self-Regulation and Arrest

The ability to control one's behavior is thought to be an important determinant of many outcomes across the life span. The study by Eisenberg et al. 2019 that we introduced in chapter 16 examined the relationship between many different aspects of self-regulation and real-world outcomes in a sample of 522 individuals tested online. In this example, we examine the question of whether a specific aspect of self-regulation known as *impulsivity* is related to the likelihood of being arrested for a crime. Impulsivity refers to the tendency to make decisions with relatively little forethought, and is generally estimated using questionnaires about one's behavioral tendencies. For example, one of these questionnaires, known as the Barratt Impulsiveness Scale, asks the person to rate their tendency to do various things (such as "act on the spur of the moment" or "plan trips well ahead of time") on a scale from 1 (rarely/never) to 4 (almost always/always). These ratings are then averaged across the many questions to compute an overall score for each person.

1: Specify Your Question of Interest

The question of interest is whether a person's level of self-reported impulsivity is related to the likelihood that they have been arrested in their life for a crime.

Table 17.1. Frequency Distribution of Number of Reported Arrests in the Eisenberg et al. Dataset

Number of arrests per individual	Number of individuals	Proportion of total
0	409	0.785
1	58	0.111
2	33	0.063
3	9	0.017
4	5	0.010
5	5	0.010
6	2	0.004

2: Identify or Collect the Appropriate Data

The study by Eisenberg et al. collected data related to different aspects of impulsivity from the following questionnaires:

- The Barratt Impulsiveness Scale, version 11 (BIS11), which includes subscales for motor impulsivity, nonplanning, and attentional impulsivity.
- The UPPS-P Impulsive Behavior Scale, where UPPS-P refers to different subscales of impulsivity: (negative) urgency, premeditation (lack of), perseverance (lack of), sensation seeking, and positive urgency.
- The Dickman Impulsivity Inventory (Dickman), which includes subscales for functional and dysfunctional impulsivity.
- The Sensation-Seeking Scale (SSS), which measures various aspects of the tendency to seek thrills or novel experiences and the tendency toward boredom.

The study also included a demographic survey and the following question: "How many times in your life have you been arrested and/or charged with illegal activities?"

3: Prepare and Visualize the Data

First, we must load the data and prepare them for analysis. The data are stored in two files, one containing the survey measures and the other containing demographic information, including the arrest question. We load these files and combine them into a single data frame.

Once the data are prepared, it's always a good idea to look at them before performing any statistical modeling, to make sure that there are no features of the data that would be problematic for the analysis plan. Let's start by looking at the distribution of arrests in the dataset. Table 17.1 shows that the majority of participants (78.5%) have never been arrested; among those who were arrested, most were arrested only once, with a very small number of individuals arrested more than two times (up to a maximum of six arrests). The

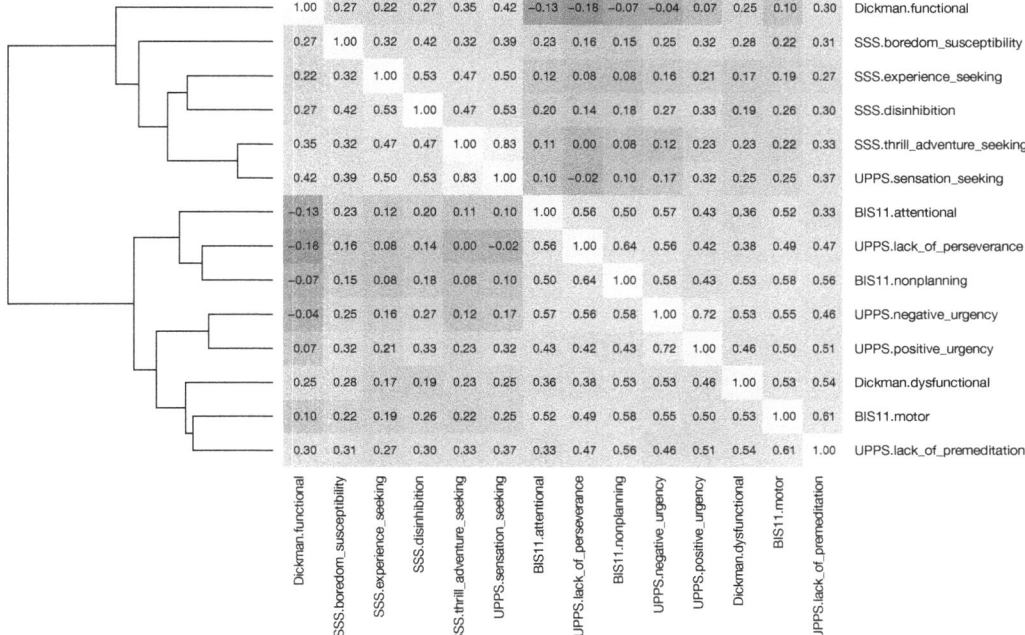

FIGURE 17.3. A heatmap of the correlation matrix of the impulsivity variables for the arrest analysis, reordered by their clustering.

shape of this distribution suggests that we probably cannot use any analyses that require assumptions of normality; this is discussed further below.

Now let's look at the impulsivity data. We have a number of different measures of impulsivity, which we would like to summarize into a smaller set of numbers for each individual. To first get an intuition about the relationship between the different variables, we can compute the correlation between each pair of variables and plot these as a heatmap, as we did in chapter 16. We can perform clustering at the same time and reorder the variables in the heatmap based on their clustering, as shown in figure 17.3.

It's clear from this figure that there are two sets of variables that seem to cluster together. The first set, at the bottom of the heatmap, are all related to classic features of impulsivity, such as lack of planning and urgency. The second set, which are at the top, are more closely related to sensation/thrill seeking and boredom susceptibility. Based on this information, we can create an average score for each individual for each of those sets of variables separately, to determine whether either is related to the likelihood of arrest. We can also standardize those variables (i.e., convert them to Z-scores), to make interpretation of the results a bit easier later on. We can visualize the resulting composite variables using a scatterplot of matrices (figure 17.4), which shows that the two resulting variables are moderately correlated and are both slightly skewed but roughly normally distributed.

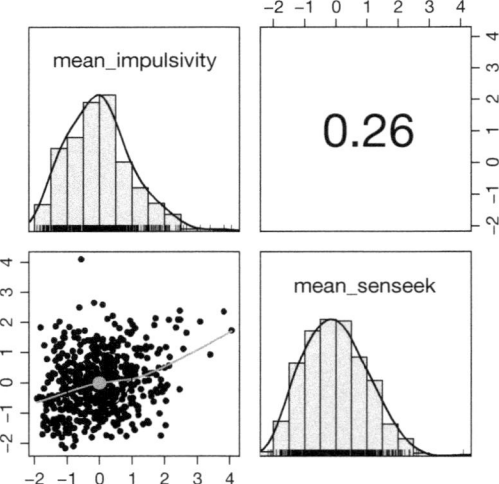

FIGURE 17.4. A scatterplot of matrices for the mean impulsivity and sensation-seeking variables derived by averaging the two clusters of original variables.

4. Determine the Appropriate Model

There are several questions that we need to ask in order to determine the appropriate statistical model for our analysis.

- What kind of dependent variable?
  We saw above that the distribution of number of arrests is highly skewed, with very few people having been arrested more than once. This means that if we were to try to fit a linear model to the number of arrests, our estimates could be highly unstable and the residuals would be highly nonnormal. When this occurs, a common approach is to *transform* the data in order to help align them better with the assumptions of the statistical modeling method. In this case, a simple alternative is to transform the arrest variable by binarizing it (that is, setting it to zero for anyone who has not been arrested and to one for anyone who has), so that we compare individuals who were arrested with those who were not.
- Are observations independent?
  Random sampling should ensure that the assumption of independence is appropriate, though there could be biases if users of the Amazon Mechanical Turk site (where these participants were recruited) are systematically different in their psychological characteristics from nonusers.
- What possible confounds exist?

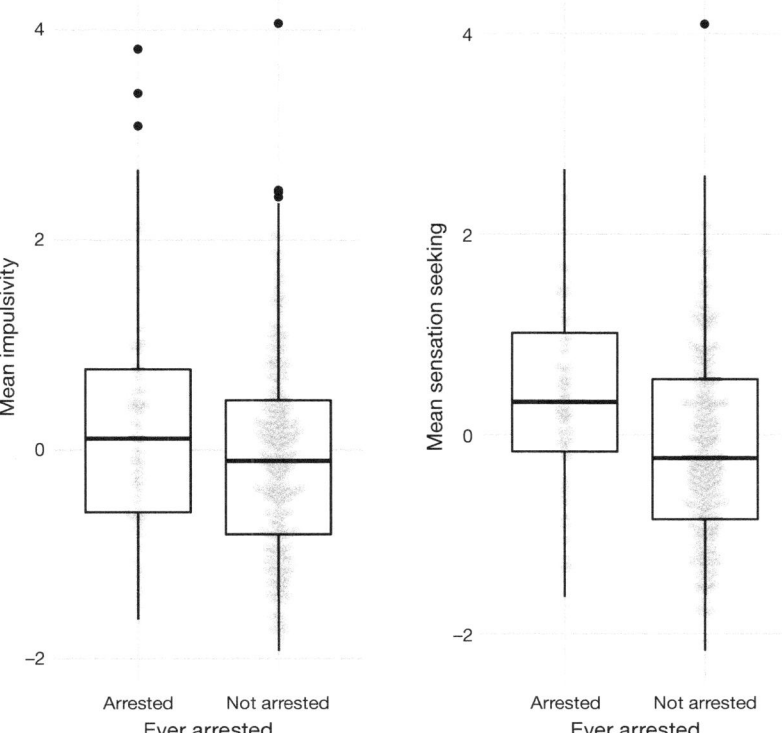

FIGURE 17.5. A plot of mean impulsivity and sensation seeking as a function of whether the individual has ever been arrested. The data are plotted using a box plot with an overlaid beehive plot that shows the distribution of the data more clearly.

A clear confound is age, since older individuals have had more chances to be arrested than younger individuals.

Another clear confound is sex, since males are known to be arrested at much higher rates than females. In this dataset, 27.4% of males had been arrested at least once, whereas only 15.6% of females had been arrested.

5. Fit the Model to The Data

First, let's look at impulsivity in relation to arrest history by using a simple approach that does not address any potential confounds. Because we have two groups (either arrested or not), we can use a two-sample *t* test to ask whether people who were arrested were more impulsive than those who were not. We start by plotting the data to get a feel for what to expect. As seen in figure 17.5, the arrested individuals appear to have higher levels of impulsivity and sensation seeking compared to the nonarrested individuals, which is confirmed by the *t* test results. Here is the *t* test output for the mean impulsivity scores:

```
##
##   Welch Two-Sample t-test
##
## data:  mean_impulsivity by everArrested
## t = 3, df = 161, p-value = 0.002
## alternative hypothesis: true difference in means between
## group Arrested and group NotArrested is greater than 0
## sample estimates:
##    mean in group Arrested   mean in group NotArrested
##    0.26                          -0.07
```

This analysis shows that impulsivity is significantly higher in those who have been arrested, with an effect size of $d = 0.33$. Let's also look at the sensation-seeking variable:

```
##
##   Welch Two-Sample t-test
##
## data:  mean_senseek by everArrested
## t = 5, df = 182, p-value = 2e-06
## alternative hypothesis: true difference in means between
## group Arrested and group NotArrested is greater than 0
## sample estimates:
##    mean in group Arrested   mean in group NotArrested
##    0.38                          -0.11
```

This also shows a significant result, with an even larger effect size of $d = 0.5$. Thus, both impulsivity and sensation seeking appear to be related to the likelihood of having been arrested.

As we mentioned above, sex and age are possible confounds for the relationship between impulsivity or sensation seeking and arrest. We would like to use a modeling approach that can allow us to investigate the relationship between these variables while also removing the potential confounding effects of age and sex. We can't do this using linear regression, since our dependent variable of interest (the binary indicator of whether a person has been arrested or not) is not normally distributed. However, we can use a related kind of model, known as a *logistic regression* model, to test this.

Logistic regression is a modification of linear regression that is meant for cases where the outcome measure (i.e., the dependent variable) is binary. Remember that in the linear regression model we predict the value of the dependent variable using a linear combination of the independent variables, multiplying each one by its regression parameter (or *beta* weight). Because of the linear nature of the equation, a unit change in the independent variable will result in the same amount of change in the predicted value of the outcome, regardless of the particular value of the independent variable. Remember the

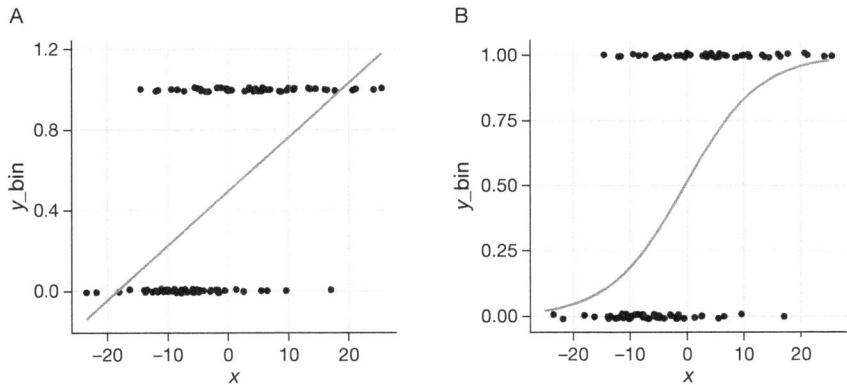

FIGURE 17.6. Plots of (A) linear regression (diagonal line) and (B) logistic regression (curve) predictions overlaid on simulated binary data. Data points are jittered in the y direction to help visualize them more clearly.

linear regression equation:

$$\hat{y}_i = \beta_0 + \beta_1 * X_i$$

where β_0 is the intercept and β_1 is the slope on variable X. Let's say that $\beta_1 = 3$; in this case, the y value increases by 3 for each additional unit of X, regardless of whether X is small or large. However, in the case of binary outcomes, this results in predicted values that do not match the binary values of zero and one and can possibly fall outside their range. We can see that in a simple example. In figure 17.6, we generate some synthetic data where a binary outcome is related to a continuous independent variable. We can see two things from the plot. First, we see that there are only two possible values of y (zero or one), such that the data points all line up at either $y = 0$ or $y = 1$, though they are jittered slightly in the plot to help see them more clearly. Second, we can see that there does seem to be a relationship between the X variable and the outcome, which is evident in the fact that there seem to be more data points on the $y = 1$ line toward the right side of the plot and more on the $y = 0$ line toward the left. Panel A shows a linear regression line that is fit to the data, and we can see the problem immediately: if we use the regression equation to predict the outcome value for an x value of 25, the predicted value is 1.17!

The intuition behind logistic regression is that, instead of predicting the actual outcome, we predict some other value that we can then link to the outcome. It is an example of what is known as a *generalized linear model*—the "generalized" part refers to the fact that the relationship between the linear predictor (i.e., the right side of the model) and the outcome variable is generalized beyond the simple linear case. In the case of logistic regression, we want to try to predict the probability of the outcome; to do this, the output of the linear predictor is related to the probability of the outcome using a *logit* function:

$$logit(p) = ln\left(\frac{p}{1-p}\right)$$

The logistic regression model for our example above thus becomes

$$ln\left(\frac{p}{1-p}\right) = \beta_0 + \beta_1 * X_i$$

You might remember the bit inside the logarithm on the left here ($\frac{p}{(1-p)}$), as it is simply the odds for probability p. It is common to say that the logistic regression equation is "linear in log odds," which means that the linear prediction from the right side of the model is related to the logarithm of the odds for the probability of the outcome, rather than the probability itself.

We can fit the logistic regression model to our example data, which results in output that looks in many ways like the output of linear regression:

```
## 
## Call:
## glm(formula = y_bin ~ x, family = binomial, data = logreg_df)
## 
## Coefficients:
##                Estimate Std. Error z value Pr(>|z|)
## (Intercept)     0.0560     0.2515    0.22     0.82
## x               0.1541     0.0336    4.58   4.6e-06 ***
## ---
## Signif. codes:  0 '***' 0.001 '**' 0.01 '*' 0.05 '.' 0.1 ' ' 1
```

We can also use this fitted model to generate the predicted values from the model, which reflect the predicted probability of y for each value of x. As shown in panel B of figure 17.6, the model predictions now fall within the same range as the data.

Now that you have a basic understanding of logistic regression, we can apply it to our impulsivity dataset. Since we have identified age and sex as potential confounding factors, we include them in our statistical model. By doing this, we ensure that any relationship between the impulsivity variables and the outcome exists even after removing the separate effects of age and sex.

Note that by dichotomizing the data as arrested versus not arrested, we are throwing away information; this may or may not be a bad thing, depending on the exact nature of the data. There are more complex generalized linear models that one could apply to this dataset, which would model both the likelihood of getting arrested or not (as our model above does) and the distribution of the number of times that someone has been arrested (known as *hurdle models*). It is recommended that you reach out to a statistical consultant if you are interested in better understanding whether this kind of model is appropriate for your dataset.

6. Perform Diagnostics on the Model to Check Assumptions

The first thing we want to do is to assess the model fit to make sure that it is appropriate. For logistic regression models, one important diagnostic is to ensure that the residuals are appropriately distributed. Remember that for linear regression we would check to see whether the residuals were normally distributed; in logistic regression, we don't expect them to be normally distributed, but we do need to check whether they have the amount of variability that is expected based on the logistic regression model. When the residual variance is larger than expected, this is known as *overdispersion*, and it is problematic because it means that the estimated standard errors will be smaller than they should be, and thus the p-values will be smaller than they should be, potentially increasing our Type I error rate. Most software packages provide the ability to test for overdispersion in the residuals of the logistic regression model; when we apply the test from our software package, we see no evidence of overdispersion:

```
##
##   Parametric dispersion test via mean Pearson-chisq statistic
##
## data:  glm_imp_arrest
## dispersion = 1, df = 515, p-value = 0.3
## alternative hypothesis: greater
```

Another assumption of logistic regression is that the predictors should have a linear relation with the log odds of the outcome; the true log odds are unknown (since for each observation we only see the binary outcome, not the underlying probability), but we can estimate them from the data using a method called *local regression* that fits a smoothed nonlinear curve to the data. Figure 17.7 plots the log odds of the observed probability (estimated from the data using local regression) against the value of each of the individual regressors in the model. For the sensation-seeking and impulsivity variables (panels B and C, respectively), we can see that the relationship is relatively linear (that is, it follows a straight line), at least in the portions of the distribution where there is a substantial amount of data. However, for the age variable (panel A), the relationship is clearly nonlinear, with an inverted U shape. We can address this by including a squared version of the age variable after subtracting its mean, which will account for the nonlinearity in the model. This is an example of how we can use the "linear" model to incorporate nonlinear effects; it remains a linear model because the nonlinear effect of Age^2 contributes to the prediction linearly (that is, by being multiplied by its regression parameter).

Finally, we should check for the potential impact of influential observations. Since we have several independent variables in our model, it is useful to examine each one separately, as shown in figure 17.7, where the size of the data points in each panel reflects the Cook's D for each observation. The low Cook's D values (all well below one) and the lack of any clear outliers on visual inspection suggests that we can proceed with hypothesis testing.

Based on our diagnostics, let's refit the model including Age^2 alongside Age:

```
##
## Call:
## glm(formula = everArrestedInt ~ mean_impulsivity + mean_senseek +
```

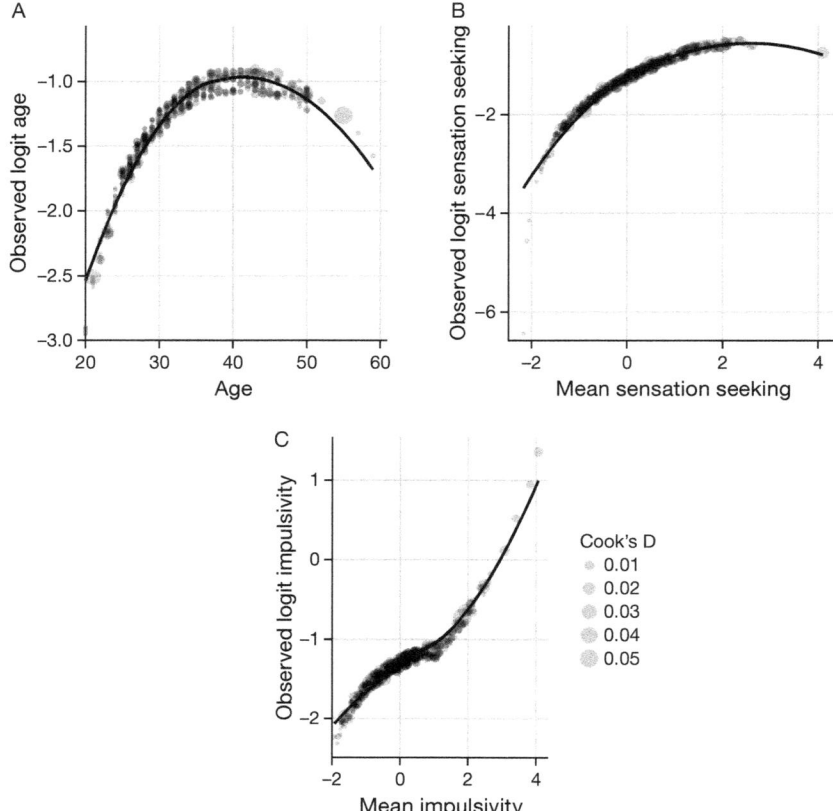

FIGURE 17.7. Plots of the three continuous variables in the model (age, mean sensation seeking, and mean impulsivity) plotted against the observed log odds of the outcome variable estimated using local regression. The size of the points in each plot is scaled by the Cook's D value for each observation.

```
##        Age + AgeSquared + Sex, family = binomial, data = impmodeldata)
##
## Coefficients:
##                   Estimate Std. Error z value Pr(>|z|)
## (Intercept)       -3.29475    0.60114   -5.48  4.2e-08 ***
## mean_impulsivity   0.30697    0.11406    2.69   0.0071 **
## mean_senseek       0.38325    0.11865    3.23   0.0012 **
## Age                0.07264    0.01864    3.90  9.8e-05 ***
## AgeSquared        -0.00423    0.00187   -2.27   0.0233 *
## SexFemale         -0.68639    0.23822   -2.88   0.0040 **
## ---
## Signif. codes:  0 '***' 0.001 '**' 0.01 '*' 0.05 '.' 0.1 ' ' 1
```

Table 17.2. Effect Sizes for Each of the Variables in the Logistic Regression Model, Expressed as Odds Ratios, Along with 95% Confidence Limits for Each Odds Ratio.

Variable	Odds ratio	2.5%	97.5%
Mean impulsivity	1.4	1.09	1.7
Mean sensation seeking	1.5	1.16	1.9
Age	1.1	1.04	1.1
Age^2	1.0	0.99	1.0
Female	0.5	0.32	0.8

7. Test Hypothesis and Quantify Effect Size

Looking at the summary of the model above, we can see that there was a significant effect (at $p < .05$) for each of the variables in the model, which means that we can reject the null hypothesis that each of the parameters is zero.

This model shows that, as expected, there is a significant relationship between a history of arrest and age and sex; the significant negative effect of Age^2 reflects the inverted U shape that we saw in the diagnostic plots above. Note that the software automatically generated a dummy variable for Sex, which represents the difference for females versus males. This means that the intercept represents the mean log odds for a (nonexistent) male with a value of zero for age, impulsivity, and sensation seeking; the `SexFemale` effect represents the offset for females (the negative value thus reflects lower log odds of arrest for a female versus a male, assuming that their age and impulsivity values are the same). Importantly, the relationship between impulsivity and sensation seeking and arrest remains even once those other effects are accounted for, suggesting that the effect is not due to those confounding factors.

We would also like to quantify the effect size for our variables of interest. Because the outcome variable for our logistic regression model is binary (having been arrested or not), we need to use a different measure of effect size: the odds ratio. This tells us how the the odds of the outcome being true change as a function of the variable of interest. We can compute these values from the estimated logistic regression parameters by simply exponentiating them, as shown in table 17.2. It's also a good idea to present the confidence intervals for each effect size estimate, so that we can see the range of parameter values that are consistent with the data.

These odds ratios provide us a way to understand how strong the effects of these variables are, but it's important to note that the different variables can't be directly compared unless they are on the same scale. The odds ratio tells us the relative change in odds given a change of one unit in the variable. For the impulsivity and sensation-seeking measures, we standardized the variables (such that the units are standard deviations), so they can both be interpreted in the same way—that is, as the change in the odds of being arrested that result from a one-standard-deviation increase in the variable. Because the age variable was not standardized, it can be interpreted in terms of its original units: the relative odds for a one-year change in age. Finally, the odds ratio for the sex variable tells us the relative odds of being arrested for a female compared to a male; in this case, the odds ratio is less than one because females have roughly half the odds of being arrested compared to males.

We would also like to know how strongly the result supports the alternative hypothesis, which we can find using the Bayes factor. For this model, we can compute the Bayes factor approximately by comparing the Bayesian information criterion (BIC) value, which is provided by our statistical software, for each of the models (following Wagenmakers 2007):

$$BF_{01} = e^{(BIC_1 - BIC_0)/2}$$

Here BF_{01} refers to the Bayes factor for hypothesis 0 (the null hypothesis) versus hypothesis 1 (the alternative hypothesis). This approximation to the Bayes factor requires that the prior probabilities for the two models (that is, the model including the impulsivity variables and the model without them) are equal, which seems reasonable in this case. The BIC values are 532.43 for the full model and 542.17 for the baseline model (remember that a lower BIC reflects a better fit). Using the BIC formula above, the Bayes factor for the baseline model compared to the full model is 0.004; since we want to know the evidence for the full model versus the baseline model, we can simply invert this (i.e., $\frac{1}{BF_{01}}$), giving a Bayes factor of 242. Based on the interpretation of Bayes factors described in chapter 11, this shows that the result provides strong evidence in favor of the hypothesis that impulsivity is related to arrest.

Example 2: Mask Wearing and Face Touching

Early in the COVID-19 pandemic there was speculation that mask wearing could have the unintended consequence of causing people to touch their face more often than when not wearing a mask; this was a point at which the mechanisms of transmission of the virus were not well understood, and it was thought that it was likely transmitted by contact, so there was great concern about face touching. This question was examined in a study performed by researchers in the Netherlands (Liebst et al. 2022), who used video from security cameras across Amsterdam captured in May/June 2020. Each person in the video was rated for whether they were wearing a mask and whether they touched their face during the period being examined. The study was *preregistered*, meaning that the design and analysis plan of the study was submitted to the Open Science Framework prior to undertaking the study. In addition, the initial study was replicated in a follow-up study. As we discuss further in chapter 18, both preregistration and replication are important tools for increasing the trustworthiness of scientific research.

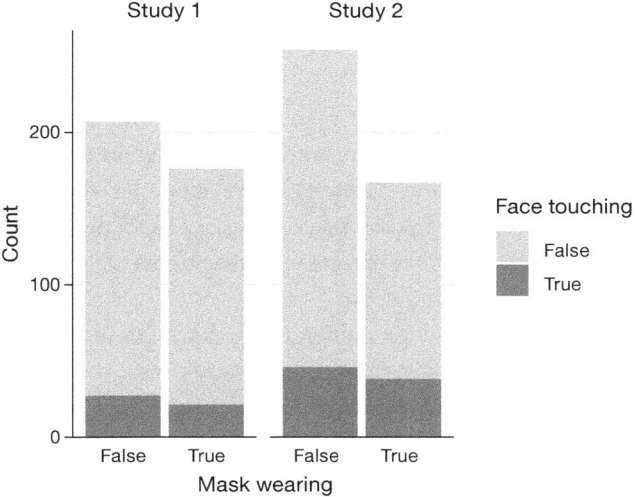

FIGURE 17.8. A stacked bar graph showing the counts of face touching versus no touching both for those who did not wear a mask for those who did in each of the two studies.

1. Specify Your Question of Interest

Are individuals wearing a face mask more likely than those not wearing a mask to touch their face?

2. Identify or Collect the Appropriate Data

The raw data for this study were shared openly via the Open Science Framework.[3] The sharing of data is another important research practice because it allows other researchers to reproduce one's findings and test other potential analyses on the data (as we are doing here).

3. Prepare and Visualize the Data

The data were cleaned according to the procedures outlined by the study authors. In particular, touches to the mask or strap were counted as touches to the face. A visual examination of the data (figure 17.8) shows that most people do not touch their face, and that there are no obvious differences in the degree of face touching depending on mask wearing.

4. Determine the Appropriate Model

- What kind of dependent variable?
  The dependent variable is the number of people who touched their face while wearing a mask or not, which is a count variable.

3. https://osf.io/7ek9d/.

We are comparing face touching across the mask-wearing and non-mask-wearing groups. Thus, a chi-squared test may be appropriate.

- Are observations independent?
  There is potential for correlation between individuals who were recorded within the same video segments, which were taken across different times and places; for example, there might be greater similarity between people who are in public at a particular time of day or particular place. Thus, we likely need to address the potential confounding effect of different video segments.
- Are there other potential confounds?
  Each individual was observed for a different amount of time that depended on how long they spent within the observation area. It seems clear that longer observation periods should result in more observed face touching, so this should also be accounted for.

  In addition, the two studies had slightly different designs, so if they are to be combined within a single analysis, then the identity of the study also needs to be included in the model.

5. Fit the Model to the Data

Let's start by performing a simple chi-squared test on data from the first study, which doesn't account for any of the potential confounds:

```
##
##  Pearson's Chi-squared test with Yates' continuity correction
##
## data:  study1$mask_wearing and study1$face_touching
## X-squared = 0.03, df = 1, p-value = 0.9
```

This shows no significant effect of mask wearing. We can perform the analogous test for the data from the second study as well:

```
##
##  Pearson's Chi-squared test with Yates' continuity correction
##
## data:  study2$mask_wearing and study2$face_touching
## X-squared = 1, df = 1, p-value = 0.3
```

Based on these analyses, it appears that there is no relationship between mask wearing and face touching. We would also like to combine the data across the two studies, as well as account for the fact that the duration of observation differed for each person. Since the outcome variable is binary, we can use the logistic regression model that we introduced in the previous example, which allows us to include additional variables as potential confounds, which in this case comprise the effect of the two different studies and the duration of observation:

```
##
## Call:
## glm(formula = face_touching ~ mask_wearing + duration_of_observation +
##     study, family = binomial, data = maskdata)
##
## Coefficients:
##                          Estimate Std. Error z value Pr(>|z|)
## (Intercept)              -2.77087    0.24823  -11.16  < 2e-16 ***
## mask_wearingTRUE          0.19190    0.19774    0.97   0.3318
## duration_of_observation   0.02899    0.00618    4.69 2.7e-06 ***
## study2                    0.57168    0.20044    2.85   0.0043 **
## ---
## Signif. codes:  0 '***' 0.001 '**' 0.01 '*' 0.05 '.' 0.1 ' ' 1
```

This result is consistent with the results of the individual chi-squared tests, showing no effect of mask wearing on face touching. It does show the expected relationship between duration of observation and face touching. However, we mentioned before that there is a potential confounding effect of the different video segments, which was not accounted for by this model.

The reason that we worry about the effect of the different video segments is that there could be clustered data, such that observations from the same segment are more similar than those from different segments. As we noted above, this is problematic for regression modeling because we generally need to assume that the residuals are independent and identically distributed, which is not the case when they are clustered. There are several different ways to address this problem, with different methods used in different research fields. Here we focus on the approach that is most common in psychology research, known as *mixed-effects modeling*; the interested reader can consult McNeish and Kelley 2019 for a thorough discussion of the differences between the various approaches to clustered data.

The concept of mixed-effects modeling relies upon a distinction between two different types of effects, known as *fixed effects* and *random effects*. There are many different ways to conceptualize this distinction, but for our purposes we consider a fixed effect as one in which we are actually interested in the different levels of the variable, and a random effect as one where we have sampled randomly from a population and we simply wish to account for the variability in the dependent variable between the different samples. Fixed effects are what we model in a standard linear regression model. In our example, we would treat mask wearing as a fixed effect, since we are explicitly interested in the difference between the two levels of the variable (wearing a mask versus not wearing one). Random effects are new here; in our example, we treat the different video segments as a random effect, which will ensure that we have properly accounted for any clustering in the data that results from differences between situations.

In this example, there are two different kinds of random effects that we could model. On the one hand, we might think that there is a different amount of face touching between the different video segments, regardless of mask wearing. We refer to this as a *random intercept* since it is akin to estimating a different baseline level of face touching for each segment, rather than a single baseline level for the entire sample. On the other hand, we might also think the effect of mask wearing on face touching is actually different across the different segments. This is referred to as a *random slope*, since we are estimating a different effect of mask wearing on face touching (which is in effect a slope for the dummy variable) for each of the unique segments. In general, we should include any random effect that is consistent with the design of the study (Barr et al. 2013); in this example, it is plausible that there could be both situational differences in the level of face touching and situational differences in the effect of mask wearing on face touching, so we include both a random intercept and a random slope.

```
## Generalized linear mixed model fit by maximum likelihood (Laplace
##    Approximation) [glmerMod]
##  Family: binomial  ( logit )
## Formula: face_touching ~ mask_wearing + duration_of_observation + (1 +
##     mask_wearing | unique_situation)
##    Data: maskdata
##
##     AIC      BIC   logLik deviance df.resid
##     704      732     -346      692      798
##
## Fixed effects:
##                         Estimate Std. Error z value Pr(>|z|)
## (Intercept)             -2.51825    0.24532  -10.27  < 2e-16 ***
## mask_wearingTRUE         0.14570    0.25382    0.57     0.57
## duration_of_observation  0.02969    0.00641    4.63  3.6e-06 ***
## ---
## Signif. codes:  0 '***' 0.001 '**' 0.01 '*' 0.05 '.' 0.1 ' ' 1
```

6. Perform Diagnostics on the Model to Check Assumptions

As in the previous example, we should first check to see whether there is any evidence of overdispersion:

```
##
##   Parametric dispersion test via mean Pearson-chisq statistic
##
## data:  glmer_result
## dispersion = 0.9, df = 798, p-value = 0.9
## alternative hypothesis: greater
```

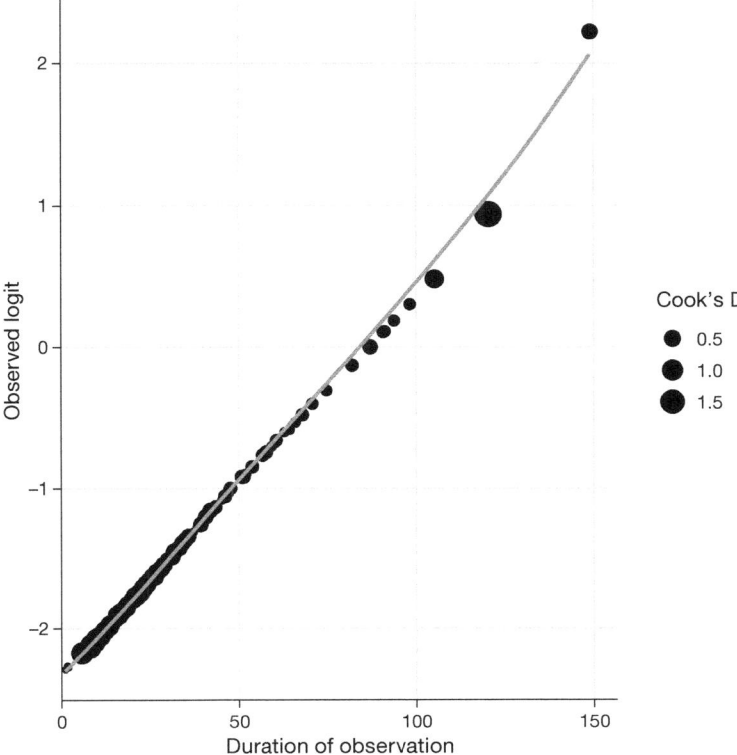

FIGURE 17.9. Diagnostic plot for mask-wearing model, showing a linear relationship between the duration of observation and the observed log odds estimated using local regression.

The test shows no evidence of overdispersion. For the one continuous independent variable (duration of observation), we can also check to confirm that it shows a linear relationship with the observed log odds in the sample. This plot (figure 17.9) shows a clear linear relationship between the independent variable and the sample logit, confirming the assumption. An examination of the Cook's D values for this dataset shows a small number of observations that have a relatively large D value (three data points with Cook's D greater than one), but the plot suggests that they are not greatly affecting the model fit (since they fall directly on the line), so we will proceed with interpretation.

7. Test Hypothesis and Quantify Effect Size

Examining the model summary above, there is no significant relationship between mask wearing and face touching in our mixed-effects model. Table 17.3 shows the estimated effect sizes (as odds ratios) and the associated confidence intervals. This shows that the data are consistent with a broad range of possible odds ratios of face touching in relation to mask wearing, from nearly half the odds to almost twice the odds.

Table 17.3. Odds Ratios and Confidence Intervals for
the Independent Variables in the Mask-Wearing Study

Independent variable	Odds ratio	2.5%	97.5%
Mask wearing	1.2	0.7	1.9
Duration of observation	1.0	1.0	1.0

Note: Odds ratios are only presented for the fixed effects; since video segments were
modeled as a random effect, that variable is not included.

Whenever we fail to reject the null hypothesis, we would like to know how strongly
the result supports the null hypothesis (rather than simply reflecting a lack of evidence),
which we can do using the Bayes factor. Using the computation from example 1 above,
we find a Bayes factor of 24.2 in favor of the null hypothesis, confirming that this result
provides relatively strong evidence for the lack of an effect of mask wearing.

Example 3: Asthma and Air Pollution

Asthma is a common health problem that is thought to relate in part to air pollution. In
this example, we combine several publicly available datasets to assess the strength of this
relationship.

1. Specify Your Question of Interest

Our hypothesis is that the level of air pollution (quantified in terms of the concentration
of fine particulate matter, known as $PM_{2.5}$) is related to the prevalence of asthma across
regions within the United States.

2. Identify or Collect the Appropriate Data

Data regarding the prevalence of asthma diagnosis were obtained from the US Centers for
Disease Control (CDC) 500 Cities Project, which collected health data from nearly 500
large US cities. The data obtained were from 2014.[4] The data are quantified at the level
of census tracts, which are geographical regions containing 1200–8000 people; there are
a total of 26,448 census tracts represented in the dataset.

Data regarding daily $PM_{2.5}$ concentrations at the census tract level for the year 2011
were also obtained from the CDC.[5] The estimated $PM_{2.5}$ level was averaged across days
within each census tract.

As you will see below, it is also important to include information regarding the relative
age and income level within each census tract, in order to address potential confounds.

4. https://chronicdata.cdc.gov/500-Cities-Places/500-Cities-Census-Tract-level-Data-GIS-Friendly-Fo
/5mtz-k78d.
5. https://data.cdc.gov/Environmental-Health-Toxicology/Daily-Census-Tract-Level-PM2-5-Concentr
ations-2011/fcqm-xrf4/data.

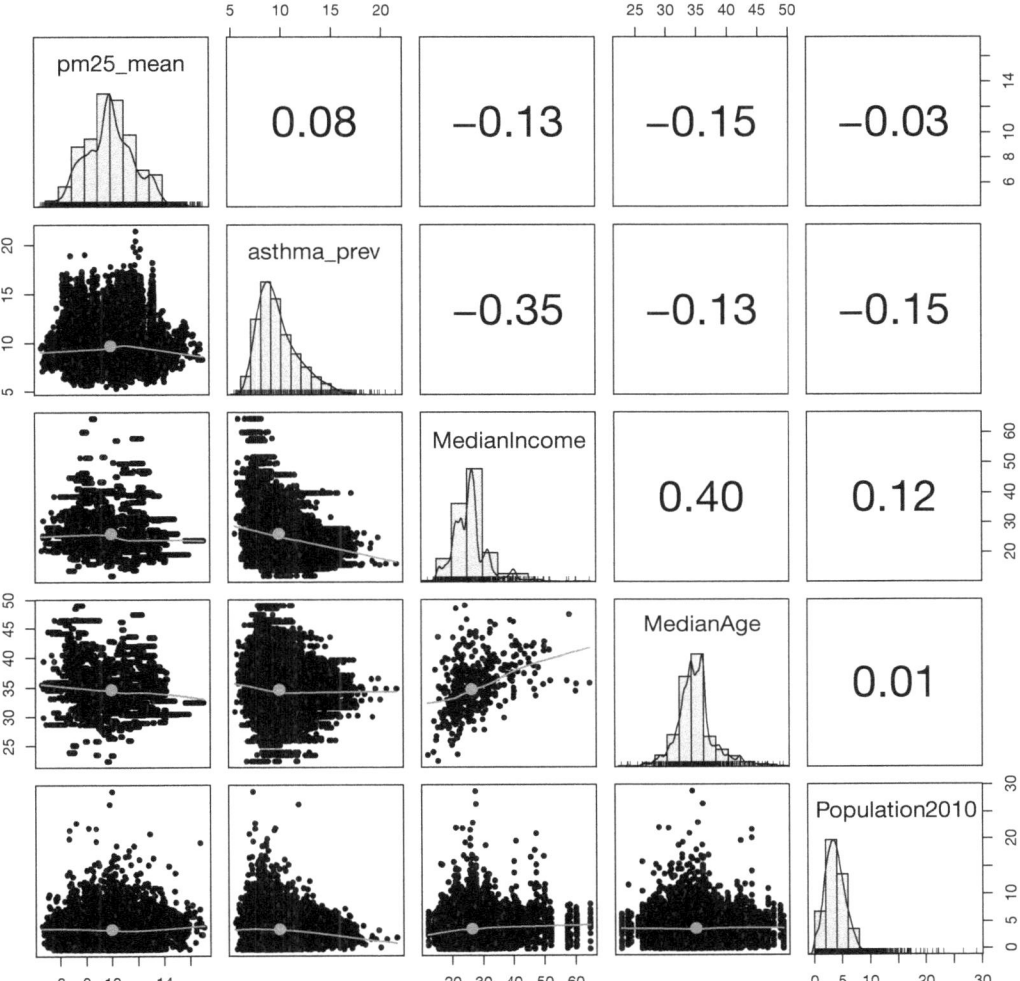

FIGURE 17.10. Scatterplot of matrices for mean annual $PM_{2.5}$ concentration values (labeled pm25_mean), along with asthma prevalence and potential confound variables. Regression lines are computed using local regression.

These data were obtained programmatically for each city using the US Census Bureau's application programming interface (API) to directly download them. One limitation of these data is that they are only available at the level of cities, rather than at the level of tracts, and thus may be missing out on variability within cities.

3. Prepare the Data for Analysis

All the data were merged into a single data frame for analysis. Initial visualization of asthma prevalence against $PM_{2.5}$ levels (figure 17.10) showed a weak correlation of 0.08 between the variables; this plot also includes some other variables that are potential confounds, including median income, median age, and population. Some of these variables are moderately skewed; we assess the potential impact of this on the results below. Note that we divide the median income and population variables by 1000 to reflect thousands of dollars and thousands of people, respectively, in order to make the parameter estimates easier to interpret; this does not change the shape of the distributions or the statistical inferences.

4. Determine the Appropriate Model

- What kind of dependent variable?
  Because the outcome variable is continuous, we start with a linear regression model.
- What are we comparing?
  We would like to assess the relation between measurements of particulate matter and asthma prevalence, both of which are continuous, so a linear regression model is appropriate.
- Are observations independent?
  Data are reported at the level of census tracts, which are selected from 494 US cities. It is likely that there will be correlations in asthma values within a city that would violate independence.
- What other possible confounds exist?
  Other potential confounds include median income and age within the city. We also include the population of the census tract, to address potential differences between smaller and larger tracts.

5. Fit the Model to the Data

We first fit a standard linear regression model to the data:

```
##
## Call:
## lm(formula = asthma_prev ~ pm25_mean + MedianIncome + MedianAge +
##     Population2010, data = pm_asthma_data)
##
## Coefficients:
##                 Estimate Std. Error t value Pr(>|t|)
## (Intercept)     12.55156    0.14979   83.80  < 2e-16 ***
## pm25_mean        0.03281    0.00620    5.29  1.2e-07 ***
## MedianIncome    -0.10958    0.00203  -53.97  < 2e-16 ***
## MedianAge        0.00470    0.00391    1.20     0.23
## Population2010  -0.11364    0.00595  -19.11  < 2e-16 ***
## ---
## Signif. codes:  0 '***' 0.001 '**' 0.01 '*' 0.05 '.' 0.1 ' ' 1
##
```

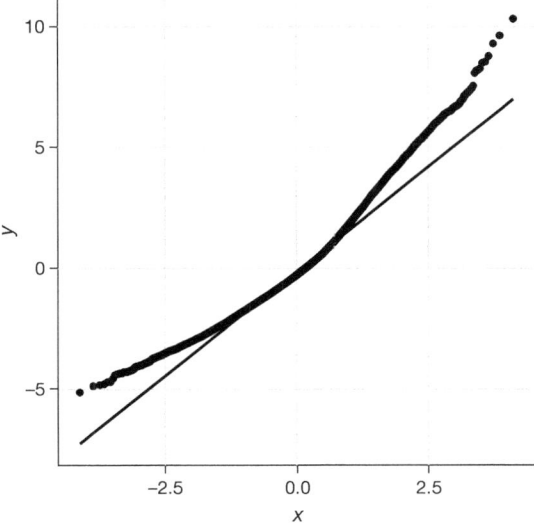

FIGURE 17.11. Q-Q plot for residuals from the simple linear regression model for asthma prevalence.

```
## Residual standard error: 1.9 on 26676 degrees of freedom
## Multiple R-squared:  0.137,  Adjusted R-squared:  0.137
## F-statistic: 1.06e+03 on 4 and 26676 DF,  p-value: <2e-16
```

This shows significant effects of each of the independent variables in the model; the very low p-values for some of these variables should be considered in light of the large sample size, which tends to make almost anything significant, as we discussed in chapter 7.

6. Perform Diagnostics on the Model to Check Assumptions

We first assess the normality of the residuals by using a Q-Q plot to identify any gross violations of the normality assumption. While there does appear to be some departure from normality in figure 17.11, the large sample size of the study should allow us to confidently interpret the hypothesis testing results despite this lack of normality.

Remember that nonnormality in the residuals can be a clue that the model is not properly specified, and here we see an example of that. In this dataset, the 26,448 observations (one for each census tract) were sampled from 494 cities in the US. If we plot residuals separately for each city (panel A of figure 17.12), we see that there are substantial differences in the distributions of residuals between cities. We can address this clustering of data within cities using a mixed-effects model that includes a random slope and intercept across cities:

```
## Linear mixed model fit by REML. t-tests use Satterthwaite's method [
## lmerModLmerTest]
```

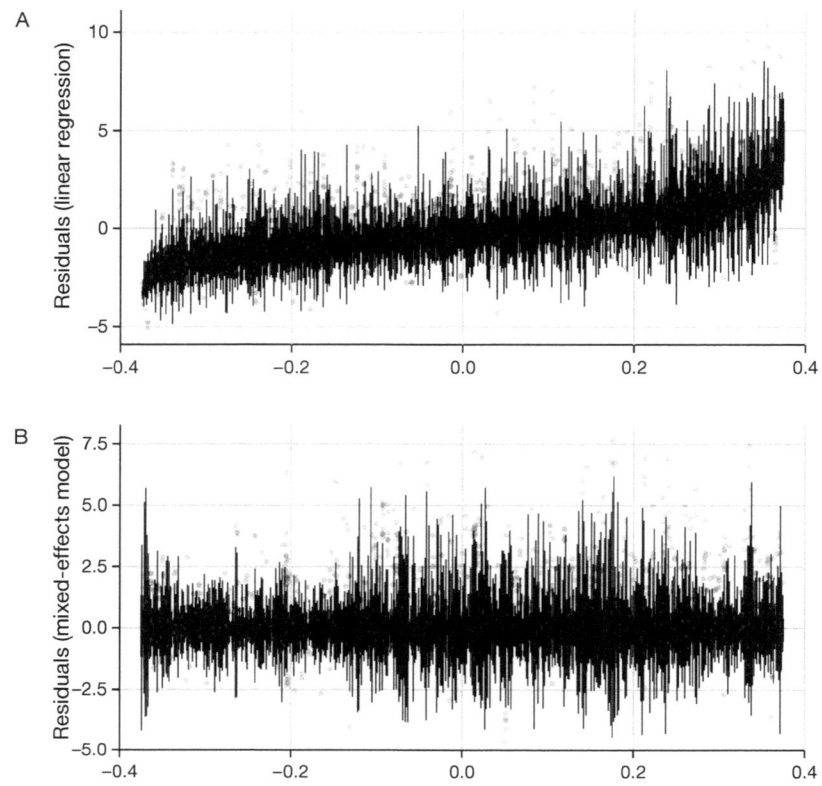

FIGURE 17.12. Box plots of residuals as a function of city for linear regression model (A) and mixed-effects model, including a random effect of city (B).

```
## Formula: asthma_prev ~ pm25_mean + MedianAge + MedianIncome + Population2010
##      (1 + pm25_mean | city)
##    Data: pm_asthma_data
##
## Fixed effects:
##                   Estimate Std. Error        df t value Pr(>|t|)
## (Intercept)       6.98e+00   1.12e+00  1.77e+02    6.22  3.6e-09 ***
## pm25_mean         3.60e-01   7.18e-02  1.49e+02    5.02  1.5e-06 ***
## MedianAge         4.97e-02   2.82e-02  1.79e+02    1.76     0.08 .
## MedianIncome     -1.03e-01   1.43e-02  1.72e+02   -7.22  1.6e-11 ***
## Population2010   -2.71e-02   4.80e-03  2.57e+04   -5.66  1.6e-08 ***
## ---
## Signif. codes:  0 '***' 0.001 '**' 0.01 '*' 0.05 '.' 0.1 ' ' 1
```

The residuals from the mixed-effects model (panel B of figure 17.12) show that residuals from all cities are now centered around zero, so we can more confidently interpret the results.

Table 17.4. Parameter Estimates and Confidence Intervals for Regression Parameters from Mixed-effects Model of Asthma Prevalence

Independent variable	Parameter estimate	2.5%	97.5%
Mean $PM_{2.5}$	0.36	0.22	0.50
Median Age	0.05	−0.01	0.10
Median Income	−0.10	−0.13	−0.07
Population (2010)	−0.03	−0.04	−0.02

7. Test Hypothesis and Quantify Effect Size

The preceding summary table from the mixed model shows that there is a significant association between air pollution and asthma prevalence, as predicted. The confidence intervals for the parameter estimates are presented in table 17.4.

This analysis shows that the data are consistent with an effect size for $PM_{2.5}$ on asthma anywhere from 0.22% to 0.50% per $\mu g/m^3$. This means that, in a population of 1000 people, each $\mu g/m^3$ of $PM_{2.5}$ is associated with roughly three to four additional asthma cases. This doesn't necessarily mean that $PM_{2.5}$ is the *cause* of those cases, but it suggests that either $PM_{2.5}$ or something related to it is playing a role. We can also use the parameter estimates to compare between the variables; for example, the effect of one $\mu g/m^3$ increase in $PM_{2.5}$ on asthma rates is roughly equivalent to a decrease in median income of about $3500.

We can also assess the effect size by computing the amount of variance in asthma prevalence that is explained by $PM_{2.5}$ levels, which is nearly 5%. This effect is both statistically significant and practically very important. By comparison, including median income in the model accounted for about 3.5% of the variance.

Example 4: Response of Plants to Nitrogen Fertilizers and Soil Tilling

The field of statistics has historically been closely tied to studies of agriculture. Ronald Fisher, one of the most influential statisticians in history, worked at the Rothamsted agricultural research station for more than a decade and spent much of his time developing methods to design agricultural experiments and analyze data regarding crop performance. In this example, we take advantage of an openly available dataset to ask how a wide range of plants respond to nitrogen fertilizer and whether this response relates to the tilling of the soil.

1. Specify Your Question of Interest

Does the effect of nitrogen fertilizer on plant growth differ for tilled versus untilled soils?

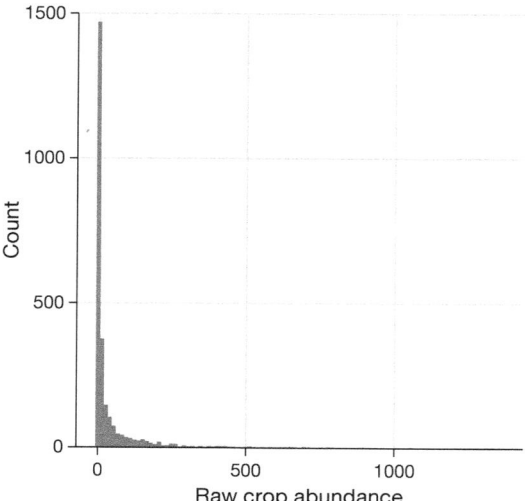

FIGURE 17.13. Histogram of raw abundance data.

2. Identify or Collect the Appropriate Data

We use a dataset from a published study of fertilizer response (Huberty, Gross, and Miller 1998), which was shared as part of a larger dataset for analyses of crop abundance (Cleland et al. 2008).

3. Prepare and Visualize the Data

The shared dataset contains data from many different experimental sites, but for this exercise we focus on a single site: the Kellogg Biological Station (KBS) located in Michigan. First, we plot the distribution of our dependent variable, which is the raw abundance of each species of plant, measured in grams per square meter. As seen in figure 17.13, this distribution appears to be long-tailed and highly skewed.

4. Determine the Appropriate Model

- What kind of dependent variable?
  The abundance measure is continuous and highly right-skewed with a very long tail. Thus, we may need to consider transforming the data.
- What are we comparing?
  We would like to assess the effect of fertilizer application in relation to tilling. Thus, an analysis of variance model is appropriate.

  The effect of interest is an interaction between fertilizer application and tilling on crop abundance.
- Are observations independent?

Data are collected across a number of different plots of land. Thus, the model needs to account for the clustering of data within a plot.

The data also include measurements of many different species and thus may exhibit clustering within species as well.

- What other possible confounds exist?

The data are collected across 10 years, and it is possible that there were overall differences between years (e.g., due to differences in weather).

5. Fit the Model to the Data

First, we fit a very simple linear model to ask whether there is an interaction between fertilization and tilling, also including the year in the model. Note that this model does not include any consideration of the different plots or different species in the data.

```
##
## Call:
## lm(formula = Rawabund ~ Fert * Experiment + Year, data = fert_data)
##
## Coefficients:
##                          Estimate Std. Error t value Pr(>|t|)
## (Intercept)              1154.521   1685.230    0.69  0.49335
## Fert1                      57.199      7.608    7.52  7.6e-14 ***
## ExperimentUntilled         -5.317      6.568   -0.81  0.41827
## Year                       -0.560      0.844   -0.66  0.50680
## Fert1:ExperimentUntilled  -35.495      9.933   -3.57  0.00036 ***
## ---
## Signif. codes:  0 '***' 0.001 '**' 0.01 '*' 0.05 '.' 0.1 ' ' 1
##
## Residual standard error: 124 on 2610 degrees of freedom
## Multiple R-squared:  0.0305, Adjusted R-squared:  0.029
## F-statistic: 20.5 on 4 and 2610 DF,  p-value: <2e-16
```

6. Perform Diagnostics on the Model to Check Assumptions

We first assess the normality of the residuals from this model. As shown in panel A of figure 17.14, there is an extreme degree of nonnormality in the residuals. Even though the sample size is large, this degree of nonnormality is concerning. It likely arises in part due to the highly skewed nature of the outcome data, so we will address it by transforming the data using a logarithmic transformation.

Another problem with this simple model is that it does not account for the fact that the dataset includes data from 132 different species of plants. If we look at the residuals separately for each species (figure 17.15), we see that they differ substantially across species,

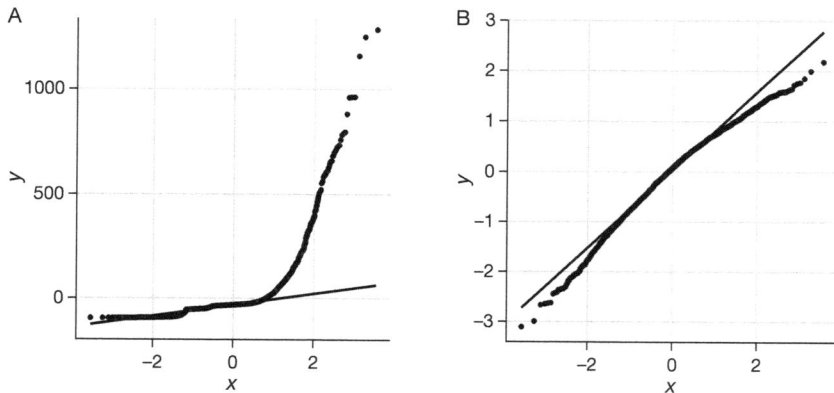

FIGURE 17.14. Q-Q plot for residuals from the simple linear regression model (A); and the mixed-effects model applied to log-transformed abundance values and including a random effect of species and plot (B).

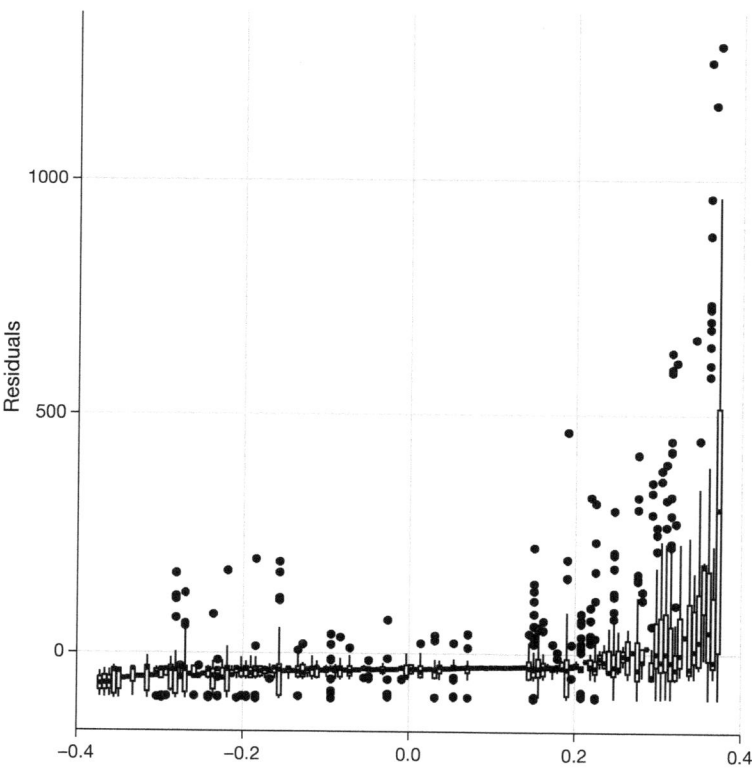

FIGURE 17.15. Residuals from simple linear model presented separately for each species of plant in the dataset, showing substantial differences in the distribution of residuals across species.

suggesting that we need to include a random effect of species in the model. To address this, we will use a mixed-effects model, as in the previous examples. This model will include random intercepts and slopes, for both the different experimental plots and the different species.

```
## Linear mixed model fit by REML. t-tests use Satterthwaite's method [
## lmerModLmerTest]
## Formula:
## log_rawabund ~ Fert * Experiment + Year + (1 + Fert | plotID_common) +
##     (1 + Fert | Species_code)
##    Data: fert_data
##
## Fixed effects:
##                          Estimate Std. Error        df t value Pr(>|t|)
## (Intercept)              1.47e+01   1.13e+01  2.55e+03    1.30  0.19268
## Fert1                    2.82e-01   7.71e-02  3.15e+01    3.66  0.00091 ***
## ExperimentUntilled       3.71e-01   9.28e-02  1.33e+01    4.00  0.00144 **
## Year                    -7.64e-03   5.65e-03  2.55e+03   -1.35  0.17650
## Fert1:ExperimentUntilled -3.26e-01  9.02e-02  2.06e+01   -3.62  0.00166 **
## ---
## Signif. codes:  0 '***' 0.001 '**' 0.01 '*' 0.05 '.' 0.1 ' ' 1
```

The inclusion of the random effects of species and plot in the model also helps reduce the nonnormality of the residuals, as shown in panel B of figure 17.14.

7. Test Hypothesis and Quantify Effect Size

The summary of the mixed-effects model presented above shows significant main effects of fertilization and tilling, as well as an interaction between fertilization and tilling, reflecting the fact that the effect of fertilization is larger for tilled versus untilled plots. Note that whenever an interaction is present, one must be careful interpreting the main effects, since the interaction shows that they are not necessarily reflective of any particular combination of conditions.

When examining an analysis of variance model with an interaction, it can be useful to look at the differences between each combination of conditions to get a better understanding of what is going on. These are often called *post hoc* tests since they are done following the identification of a significant interaction. However, we have to adjust the p-values for each test to account for the fact that we are doing multiple tests. We use a particular method called the *Tukey adjustment* to help control our Type I error:

```
## contrast                     estimate   SE df t.ratio p.value
## Fert0 Tilled - Fert1 Tilled     -0.28 0.079 33  -3.600  0.0100
## Fert0 Tilled - Fert0 Untilled   -0.37 0.093 14  -4.000  0.0100
```

```
## Fert0 Tilled - Fert1 Untilled      -0.33 0.093 28  -3.500  0.0100
## Fert1 Tilled - Fert0 Untilled      -0.09 0.100 36  -0.900  0.8100
## Fert1 Tilled - Fert1 Untilled      -0.05 0.090 23  -0.500  0.9600
## Fert0 Untilled - Fert1 Untilled     0.04 0.065 19   0.700  0.9100
##
## degrees-of-freedom method: kenward-roger
## p-value adjustment: Tukey method for comparing a family of four estimates
```

Because we transformed the data prior to analysis, we can't interpret the parameter estimates in terms of the original units of the data. However, because we converted the transformed abundance scores into Z-scores, we can interpret the parameter estimates in terms of standard deviations of the transformed data. This shows that untilled plots had roughly a one-third standard deviation higher log abundance regardless of fertilization, and that fertilization increased log abundance for tilled plots by a bit less than a one-third standard deviation. It appears that the positive effect of leaving the land untilled was roughly the same as the effect of fertilizer application.

We can also examine the amount of variance accounted for by each of the factors in the model. The inclusion of the interaction between fertilizing and tilling accounted for about 1.1% more variance than the model without an interaction, and the full model accounted for about 2.1% more variance than a baseline model that only included the effect of the year along with the random effects of plot and species. Thus, while significant, these factors accounted for relatively little variance in the data.

Getting Help

Whenever you are analyzing real data, it's useful to check your analysis plan with a trained statistician, as there are many potential problems that could arise in real data. In fact, it's best to speak to a statistician before you even start the project, as their advice regarding the design or implementation of the study could save you major headaches down the road. Most universities have statistical consulting offices that offer free assistance to members of the university community. Understanding the content of this book won't prevent you from needing their help at some point, but it will help you have a more informed conversation with them and better understand the advice that they offer.

Problems

1. Describe the set of steps that one should take in analyzing a dataset to test a scientific hypothesis.
2. When is the assumption of normally distributed residuals particularly important for a linear regression model?
3. A researcher has a dataset in which he measured whether or not a set of individuals could run one mile in under 10 minutes, based on measurements of muscle strength

and oxygen capacity. Would this be more appropriate for linear regression or logistic regression, and why?

4. Which of the following is true about outliers? Choose all that apply.
   - They should always be removed.
   - They can sometimes reflect the true nature of the data.
   - They can sometimes reflect errors in measurement.
   - They require a scientific understanding in order to determine how to address them.

5. Describe the concept of collider bias and how it relates to the specification of a statistical model.

6. Describe the relationship between logistic and linear regression.

7. A researcher runs two statistical models, which result in BIC values of 327 for the full model and 385 for the null model. Compute the Bayes factor (BF_{01}) and describe your interpretation of the result.

8. What is the main application of mixed-effects models described in the chapter?
   - To combine data across multiple individuals
   - To address the presence of clustered data due to random effects
   - To allow the mixing of normally distributed and binary data
   - To address the problem of outliers

9. Which of the following are potential causes of nonnormality in the residuals of a linear model? Choose all that apply.
   - The model is not properly specified.
   - There is structure in the data that has not been appropriately modeled.
   - The outcome variable is highly skewed.

10. What kind of mathematical transformation might one apply to a variable that is skewed in order to increase the normality of the residuals of a linear model?

18

Doing Reproducible Research

Most people think that science is a reliable way to answer questions about the world. When our physician prescribes a treatment, we trust that it has been shown to be effective through research, and we have similar faith that the airplanes we fly in aren't going to fall from the sky. However, since 2005 there has been an increasing concern that science may not always work as well as we have long thought it does. In this chapter, we discuss concerns about the reproducibility of scientific research, and outline the steps that we can take to make sure our statistical results are as reproducible as possible.

Learning Objectives

Having read this chapter, you should be able to

- Describe the concept of p-hacking and its effects on scientific practice.
- Describe the concept of positive predictive value and its relationship to statistical power.
- Describe the concept of preregistration and how it can help protect against questionable research practices.

How We Think Science Should Work

Let's say that we are interested in a research project on how children choose what to eat. This is a question that was asked in a study by the well-known eating researcher Brian Wansink and his colleagues in 2012. The standard (and, as we will see, somewhat naive) view goes something like this:

- You start with a hypothesis.
  Branding with popular media characters should cause children to choose "healthy" food more often.
- You collect some data.
  Offer children the choice between a cookie and an apple with either an Elmo-branded sticker or a control sticker, and record what they choose.

- You do statistics to test the null hypothesis.

  "The preplanned comparison shows Elmo-branded apples were associated with an increase in a child's selection of an apple over a cookie, from 20.7% to 33.8% ($\chi^2 = 5.158$; $P = .02$)" (Wansink, Just, and Payne 2012).

- You make a conclusion based on the data.

  "This study suggests that the use of branding or appealing branded characters may benefit healthier foods more than they benefit indulgent, more highly processed foods. Just as attractive names have been shown to increase the selection of healthier foods in school lunchrooms, brands and cartoon characters could do the same with young children"(Wansink, Just, and Payne 2012).

How Science (Sometimes) Actually Works

Brian Wansink is well known for his book *Mindless Eating*, and his fee for corporate speaking engagements was at one point in the tens of thousands of dollars. In 2017, a group of researchers began to scrutinize some of his published findings, starting with a set of papers about how much pizza people ate at a buffet. The researchers asked Wansink to share the data from the studies, but he refused; so they dug into his published work and found a large number of inconsistencies and statistical problems in the papers. The publicity around this analysis led a number of others to dig into Wansink's past, including obtaining emails between Wansink and his collaborators. As reported by Stephanie Lee at Buzzfeed, these emails showed just how far Wansink's actual research practices were from the naive model:

> Back in September 2008, when Payne was looking over the data soon after it had been collected, he found no strong apples-and-Elmo link—at least not yet. . . . "I have attached some initial results of the kid study to this message for your report," Payne wrote to his collaborators. "Do not despair. It looks like stickers on fruit may work (with a bit more wizardry)." . . . Wansink also acknowledged the paper was weak as he was preparing to submit it to journals. The p-value was 0.06, just shy of the gold standard cutoff of 0.05. It was a "sticking point," as he put it in a Jan. 7, 2012, email. . . . "It seems to me it should be lower," he wrote, attaching a draft. "Do you want to take a look at it and see what you think. If you can get the data, and it needs some tweaking, it would be good to get that one value below .05." . . . Later in 2012, the study appeared in the prestigious *JAMA Pediatrics*, the 0.06 p-value intact. But in September 2017, it was retracted and replaced with a version that listed a p-value of 0.02. And a month later, it was retracted yet again for an entirely different reason: Wansink admitted that the experiment had not been done on 8- to 11-year-olds, as he'd originally claimed, but on preschoolers. (Lee 2018)

This kind of behavior finally caught up with Wansink; at least eighteen of his research studies have been retracted, and in 2018 he resigned from his faculty position at Cornell University.

The Reproducibility Crisis in Science

While we think that the kind of fraudulent behavior seen in Wansink's case is relatively rare, it has become increasingly clear that problems with reproducibility are much more widespread in science than previously thought. This became particularly evident in 2015, when a large group of researchers published a study in the journal *Science* titled "Estimating the Reproducibility of Psychological Science"(Open Science Collaboration 2015). In this study, known as the Reproducibility Project: Psychology, the researchers took 100 published studies in psychology and attempted to reproduce the results originally reported in the papers. Their findings were shocking: whereas 97% of the original papers had reported statistically significant findings, only 37% of these effects were statistically significant in the replication study. Although these problems in psychology have received a great deal of attention, they seem to be present in nearly every area of science, from cancer biology (Errington et al. 2014) and chemistry (Baker 2017) to economics (Christensen and Miguel 2016) and the social sciences (Camerer et al. 2018).

The reproducibility crisis that emerged after 2010 was actually predicted by John Ioannidis, a physician from Stanford who wrote a paper in 2005 titled "Why Most Published Research Findings Are False"(Ioannidis 2005). In this article, Ioannidis argued that the use of null hypothesis statistical testing in the context of modern science necessarily leads to high levels of false results.

Positive Predictive Value and Statistical Significance

Ioannidis's analysis focused on a concept known as *positive predictive value*, which is defined as the proportion of positive results (generally translated to "statistically significant findings") that are true:

$$PPV = \frac{p(\textit{true positive result})}{p(\textit{true positive result}) + p(\textit{false positive result})}$$

Assuming that we know the probability that our hypothesis is true $(p(\textit{hIsTrue}))$, then the probability of a true positive result is simply $p(\textit{hIsTrue})$ multiplied by the statistical power of the study:

$$p(\textit{true positive result}) = p(\textit{hIsTrue}) * (1 - \beta)$$

where β is the false negative rate. The probability of a false positive result is determined by $p(\textit{hIsTrue})$ and the false positive rate α:

$$p(\textit{false positive result}) = (1 - p(\textit{hIsTrue})) * \alpha$$

PPV is then defined as

$$PPV = \frac{p(hIsTrue) * (1 - \beta)}{p(hIsTrue) * (1 - \beta) + (1 - p(hIsTrue)) * \alpha}$$

Let's first take an example where the probability of our hypothesis being true is high—say, 0.8—though note that in general we cannot actually know this probability. Let's say that we perform a study with the standard values of $\alpha = 0.05$ and $\beta = 0.2$. We can compute the PPV as

$$PPV = \frac{0.8 * (1 - 0.2)}{0.8 * (1 - 0.2) + (1 - 0.8) * 0.05} = 0.98$$

This means that if we find a positive result in a study where the hypothesis is likely to be true and power is high, then its likelihood of being true is high. Note, however, that a research field where hypotheses have such a high likelihood of being true is probably not a very interesting field of research; research is most important when it tells us something unexpected!

Let's do the same analysis for a field where $p(hIsTrue) = 0.1$—that is, most of the hypotheses being tested are false. In this case, PPV is

$$PPV = \frac{0.1 * (1 - 0.2)}{0.1 * (1 - 0.2) + (1 - 0.1) * 0.05} = 0.307$$

This means that in a field where most of the hypotheses are likely to be wrong (that is, an interesting scientific field where researchers are testing risky hypotheses), even when we find a positive result it is more likely to be false than true! In fact, this is just another example of the base rate effect that we discussed in chapter 6—when an outcome is unlikely, then it's almost certain that most positive outcomes will be false positives. We can simulate this to show how PPV relates to statistical power, as a function of the prior probability of the hypothesis being true (figure 18.1). Unfortunately, statistical power remains low in many areas of science (Smaldino and McElreath 2016), suggesting that many published research findings are false. An amusing example of this was seen in a paper by Jonathan Schoenfeld and John Ioannidis (2013), titled "Is Everything We Eat Associated with Cancer? A Systematic Cookbook Review." They examined a large number of papers that had assessed the relationship between different foods and cancer risk, and found that 80% of ingredients had been associated with either increased or decreased cancer risk. In most of these cases, the statistical evidence was weak, and when the results were combined across studies, the result was null.

The Winner's Curse

Another kind of error can also occur when statistical power is low: our estimates of the effect size will be inflated. This phenomenon often goes by the term *winner's curse*, which comes from economics, where it refers to the fact that for certain types of auctions (where

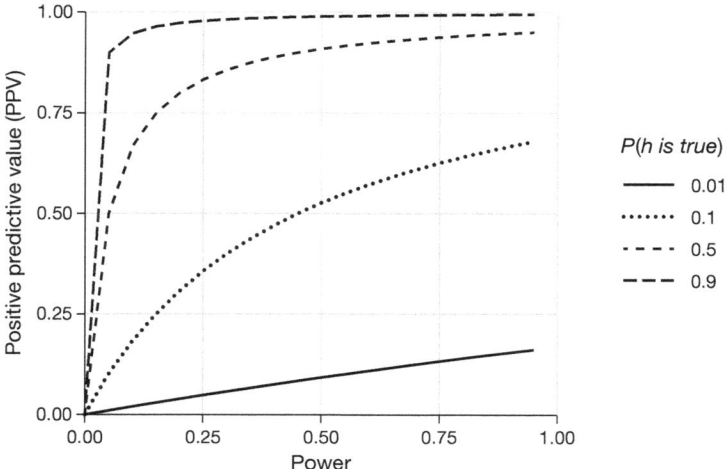

FIGURE 18.1. A simulation of positive predictive value as a function of statistical power (plotted on the x-axis) and prior probability of the hypothesis being true (plotted as separate lines).

the value is the same for everyone, like a jar of quarters, and the bids are private), the winner is guaranteed to pay more than the good is worth. In science, the winner's curse refers to the fact that the effect size estimated from a significant result (i.e., a winner) is almost always an overestimate of the true effect size.

We can simulate this in order to see how the estimated effect size for significant results is related to the actual underlying effect size. Let's generate data for which there is a true effect size of $d = 0.2$, and estimate the effect size for those results where there is a significant effect detected. Panel A of figure 18.2 shows that, when power is low, the estimated effect size for significant results can be highly inflated compared to the actual effect size.

We can look at a single simulation to see why this is the case. In panel B of figure 18.2, you can see a histogram of the estimated effect sizes for 1000 samples, separated by whether the test was statistically significant. It should be clear from the figure that, if we estimate the effect size only based on significant results, our estimate will be inflated; only when most results are significant (i.e., power is high and the effect is relatively large) will our estimate come near the actual effect size.

Questionable Research Practices

A popular book entitled *The Compleat Academic: A Career Guide*, published by the American Psychological Association (Darley, Zanna, and Roediger, eds. 2004), aims to provide aspiring researchers with guidance on how to build a career. In a chapter by well-known social psychologist Daryl Bem titled "Writing the Empirical Journal Article," Bem provides some suggestions about how to write a research paper. Unfortunately, the practices that he suggests are deeply problematic and have come to be known as *questionable research practices* (QRPs).

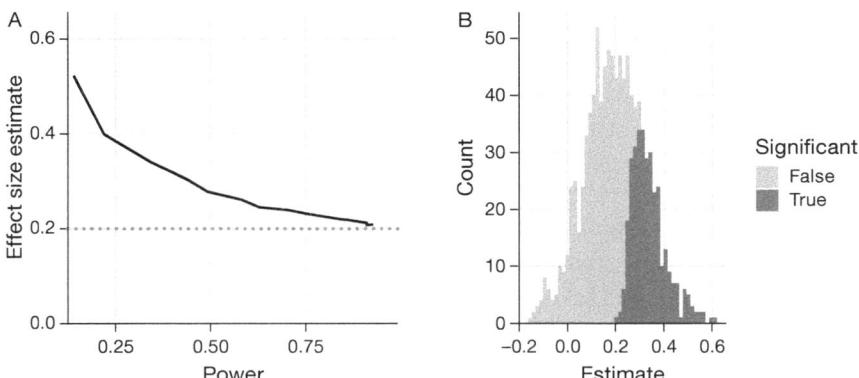

FIGURE 18.2. (A) A simulation of the winner's curse as a function of statistical power (x-axis). The solid line shows the estimated effect size, and the dotted line shows the actual effect size. (B) A histogram showing effect size estimates for a number of samples from a dataset, with significant results shown in dark gray and nonsignificant results in light gray.

Which article should you write? There are two possible articles you can write: (1) the article you planned to write when you designed your study or (2) the article that makes the most sense now that you have seen the results. They are rarely the same, and the correct answer is (2). (Bem 2004, p. 186)

What Bem suggests here is known as *HARKing* (hypothesizing after the results are known)(Kerr 1998). This might seem innocuous, but it is problematic because it allows the researcher to reframe a post hoc conclusion (which we should take with a grain of salt) as an a priori prediction (in which we would have stronger faith). In essence, it allows the researcher to rewrite their theory based on the facts, rather that using the theory to make predictions and then test them—akin to moving the goalpost so that it ends up wherever the ball goes. It thus becomes very difficult to disconfirm incorrect ideas, since the goalpost can always be moved to match the data. Bem continues:

Analyzing data Examine them from every angle. Analyze the sexes separately. Make up new composite indices. If a datum suggests a new hypothesis, try to find further evidence for it elsewhere in the data. If you see dim traces of interesting patterns, try to reorganize the data to bring them into bolder relief. If there are participants you don't like, or trials, observers, or interviewers who gave you anomalous results, drop them (temporarily). Go on a fishing expedition for something—anything—interesting. No, this is not immoral. (Bem 2004, p. 186)

What Bem suggests here is known as *p-hacking*, which refers to trying many different analyses until one finds a significant result. Bem is correct that, if one were to report every analysis done on the data, this approach would not be "immoral." However, it is rare to see a paper discuss all of the analyses that were performed on a dataset; rather, papers often only present the analyses that *worked*—which usually means that they found a statistically significant result. There are many different ways that one might p-hack:

- Analyze data after every subject and stop collecting data once p<.05.
- Analyze many different variables but report only those with p<.05.
- Collect many different experimental conditions but report only those with p<.05.
- Exclude participants to get p<.05.
- Transform the data to get p<.05.

A well-known paper by Simmons, Nelson, and Simonsohn (2011) showed that the use of these kinds of p-hacking strategies could greatly increase the actual false positive rate, resulting in a high number of false positive results.

ESP or QRP?

In 2011, that same Daryl Bem published an article that claimed to have found scientific evidence for extrasensory perception:

> This article reports 9 experiments, involving more than 1,000 participants, that test for retroactive influence by "time-reversing" well-established psychological effects so that the individual's responses are obtained before the putatively causal stimulus events occur.... The mean effect size (d) in psi performance across all 9 experiments was 0.22, and all but one of the experiments yielded statistically significant results. (Bem 2011, p. 407)

As researchers began to examine Bem's article, it became clear that he had engaged in all of the QRPs that he had recommended in the chapter discussed above. As Tal Yarkoni pointed out in a blog post that examined the article:[1]

- Sample sizes varied across studies.
- Different studies appear to have been lumped together or split apart.
- The studies allow many different hypotheses, and it's not clear which were planned in advance.
- Bem used one-tailed tests even when it's not clear that there was a directional prediction (so alpha is really 0.1).
- Most of the p-values are very close to 0.05.
- It's not clear how many other studies were run but not reported.

Doing Reproducible Research

In the years since the reproducibility crisis arose, there has been a robust movement to develop tools to help protect the reproducibility of scientific research.

---

1. http://www.talyarkoni.org/blog/2011/01/10/the-psychology-of-parapsychology-or-why-good-rese archers-publishing-good-articles-in-good-journals-can-still-get-it-totally-wrong/.

Preregistration

One of the ideas that has gained the greatest traction is *preregistration*, in which one submits a detailed description of a study (including all data analyses) to a trusted repository (such as the Open Science Framework or AsPredicted.org). By specifying one's plans in detail prior to analyzing the data, preregistration provides greater faith that the analyses do not suffer from p-hacking or other questionable research practices.

The effects of preregistration in clinical trials in medicine have been striking. In 2000, the National Heart, Lung, and Blood Institute (NHLBI) began requiring all clinical trials to be preregistered using the system at ClinicalTrials.gov. This provides a natural experiment to observe the effects of study preregistration. When Kaplan and Irvin (2015) examined clinical trial outcomes over time, they found that the number of positive outcomes in clinical trials was greatly reduced after 2000 compared to before. While there are many possible causes, it seems likely that prior to study registration researchers were able to change their methods or hypotheses in order to find a positive result, which became more difficult after registration was required.

Reproducible Practices

The paper by Simmons, Nelson, and Simonsohn (2011) lays out a set of suggested practices for making research more reproducible, all of which should become standard for researchers:

- Authors must decide the rule for terminating data collection before data collection begins and report this rule in the article.
- Authors must collect at least 20 observations per cell or else provide a compelling cost-of-data-collection justification.
- Authors must list all variables collected in a study.
- Authors must report all experimental conditions, including failed manipulations.
- If observations are eliminated, authors must also report what the statistical results are if those observations are included.
- If an analysis includes a covariate, authors must report the statistical results of the analysis without the covariate.

Replication

One of the hallmarks of science is the idea of *replication*—that is, other researchers should be able to perform the same study and obtain the same result. Unfortunately, as we saw in the outcome of the Reproducibility Project discussed earlier in the chapter, many findings are not replicable. The best way to ensure replicability of your research is to first replicate it on your own; for some studies this won't be possible, but whenever it is possible, you should make sure that your finding holds up in a new sample. That new sample should be sufficiently powered to find the effect size of interest; in many cases, this will actually require a larger sample than the original.

It's important to keep a couple of things in mind with regard to replication. First, the fact that a replication attempt fails does not necessarily mean that the original finding was false; remember that, with the standard level of 80% power, there is still a one in five chance that the result will be nonsignificant, even if there is a true effect. For this reason, we generally want to see multiple replications of any important finding before we decide whether or not to believe it, and we generally want the replication attempts to have higher levels of power than the original. Unfortunately, many fields, including psychology, have failed to follow this advice in the past, leading to "textbook" findings that turn out to be likely false. With regard to Daryl Bem's studies of ESP, a large replication attempt involving seven studies failed to replicate his findings (Galak et al. 2012).

Second, remember that the p-value doesn't provide us with a measure of the likelihood of a finding to replicate. As we discussed previously, the p-value is a statement about the likelihood of one's data under a specific null hypothesis; it doesn't tell us anything about the probability that the finding is actually true (as we learned in chapter 11). In order to know the likelihood of replication, we need to know the probability that the finding is true, which we generally don't know.

Doing Reproducible Data Analysis

So far we have focused on the ability to replicate other researchers' findings in new experiments, but another important aspect of reproducibility is to be able to reproduce someone's analyses on their own data, which we refer to as *computational reproducibility*. This requires that researchers share both their data and their analysis code, so that other researchers can both try to reproduce the result as well as potentially test different analysis methods on the same data. There is an increasing move in psychology toward open sharing of code and data; for example, the journal *Psychological Science* now provides "badges" to papers that share research materials, data, and code, as well as for pre-registration.

The ability to reproduce analyses is one reason that we strongly advocate for the use of scripted analyses (such as those using R) rather than using a "point-and-click" software package. It's also a reason that we advocate for the use of free and open source software (like R) as opposed to commercial software packages, which would require others to buy the software in order to be able to reproduce any analyses.

There are many ways to share both code and data. A common way to share code is via websites that support *version control* for software, such as Github. Small datasets can also be shared via these same sites; larger datasets can be shared through data-sharing portals, such as Zenodo, or through specialized portals for specific types of data, such as OpenNeuro for neuroimaging.

Conclusion: Doing Better Science

It is every scientist's responsibility to improve their research practices in order to increase the reproducibility of their research. It is essential to remember that the goal of research is not to find a significant result; rather, it is to ask and answer questions about nature in the most truthful way possible. Most of our hypotheses will be wrong, and we should be comfortable with that, so that when we find one that's right, we will be even more confident in its truth.

Suggested Reading

- *Rigor Mortis: How Sloppy Science Creates Worthless Cures, Crushes Hope, and Wastes Billions*, by Richard Harris. An outstanding overview of the reproducibility crisis in science.
- *Statistics Done Wrong: The Woefully Complete Guide*, by Alex Reinhart. A humorous guide to all the ways that statistics can be misused.

Problems

1. What percentage of published psychology studies were found to be replicable in the 2015 study by the Reproducibility Project: Psychology?
   - 18%
   - 37%
   - 51%
   - 86%
2. What did John Ioannidis show in his 2005 paper titled "Why Most Published Research Findings Are False"? Choose all that apply.
   - Most researchers made up their data.
   - Statistical power is related to the likelihood that a positive result is actually true.
   - Smaller studies are more likely to have Type I errors.
   - A positive result from a large study is more likely to be true than a positive result from a small study.
3. Ten different researchers perform independent studies to determine whether peanut allergies are related to the mother's intake of peanuts during pregnancy. Six of the studies find a statistically significant relationship, with effect sizes ranging from $d = 0.2$ to $d = 0.5$. The concept of the winner's curse tells us that the true effect size is likely (larger, small, identical to) the largest effect size ($d = 0.5$) in the set of studies.
4. Describe the concept of HARKing and why it is problematic for research quality.
5. Which of the following would be examples of p-hacking? Choose all that apply.
   - Trying five different analysis methods on the same data and reporting only the most significant result

- Performing 1,000,000 hypothesis tests and reporting the results after correcting for multiple tests
- Using cross-validation to determine the predictive performance of a model
- Analyzing the data after each new observation and stopping the study when the result becomes significant

6. Which of the following are included in the suggested rules for improving reproducibility by Simmons and colleagues? Choose all that apply.
   - Authors must collect at least 50 observations per group or provide a compelling justification for not doing so.
   - Authors must report all financial conflicts of interest.
   - Authors must report all experimental conditions, including those that failed.
   - Authors must decide the rule for terminating data collection before data collection begins and report this rule in the article.

BIBLIOGRAPHY

Bakan, David (1966). "The test of significance in psychological research." In: *Psychol Bull* 66.6, pp. 423–37.

Baker, Monya (2017). "Reproducibility: Check your chemistry." In: *Nature* 548.7668 (Aug.), pp. 485–88. https://doi.org/10.1038/548485a.

Barr, Dale J et al. (2013). "Random effects structure for confirmatory hypothesis testing: Keep it maximal." In: *J Mem Lang* 68.3, pp. 255–78. https://doi.org/10.1016/j.jml.2012.11.001.

Bem, Daryl J (2004). "Writing the empirical journal article." In: J M Darley, M P Zanna, and H L Roediger (eds.), *The Compleat Academic: A Career Guide*, pp. 185–219. Washington, DC: American Psychological Association.

Bem, Daryl J (2011). "Feeling the future: Experimental evidence for anomalous retroactive influences on cognition and affect." In: *J Pers Soc Psychol* 100.3, pp. 407–25. https://doi.org/10.1037/a0021524.

Benjamin, Daniel J et al. (2018). "Redefine Statistical Significance." In: *Nat Hum Behav* 2.1, pp. 6–10.

Breiman, Leo (2001). "Statistical modeling: The two cultures (with comments and a rejoinder by the author)." In: *Statist Sci* 16.3 (Aug.) Pp. 199–231. https://doi.org/10.1214/ss/1009213726.

Camerer, Colin F et al. (2018). "Evaluating the replicability of social science experiments in *Nature* and *Science* between 2010 and 2015." In: *Nat Hum Behav* 2, pp. 637–44.

Christensen, Garret S and Edward Miguel (2016). "Transparency, reproducibility, and the credibility of economics research." Working paper 22989. National Bureau of Economic Research. http://www.nber.org/papers/w22989.

Cleland, Elsa E et al. (2008). "Species responses to nitrogen fertilization in herbaceous plant communities, and associated species traits." In: *Ecology* 89.4, p. 1175. https://doi.org/10.1890/07-1104.1.

Copas, J B (1983). "Regression, prediction and shrinkage (with discussion)." In: *J R Statist Soc B* 45, pp. 311–54.

Darley, John M, Mark P Zanna, and Henry L Roediger, eds. (2004). *The Compleat Academic: A Career Guide*. 2nd ed. Washington, DC: American Psychological Association.

Dehghan, Mahshid et al. (2017). "Associations of fats and carbohydrate intake with cardiovascular disease and mortality in 18 countries from five continents (PURE): A prospective cohort study." In: *Lancet* 390.10107, pp. 2050–62. https://doi.org/10.1016/S0140-6736(17)32252-3.

Efron, Bradley (1998). "R. A. Fisher in the 21st century." Paper presented at the 1996 R. A. Fisher Lecture. In: *Statist. Sci.* 13.2 (May), pp. 95–122. https://doi.org/10.1214/ss/1028905930.

Einstein, Albert (1934). "On the method of theoretical physics." In: *Phil Sci* 1.2 (Apr.), pp. 163–69.

Eisenberg, Ian W et al. (2019). "Uncovering the structure of self-regulation through data-driven ontology discovery." In: *Nat Commun* 10.1 (May), p. 2319. https://doi.org/10.1038/s41467-019-10301-1.

Enders, Craig K (2022). *Applied Missing Data Analysis*. 2nd ed. New York: Guilford Press.

Errington, Timothy M et al. (2014). "An open investigation of the reproducibility of cancer biology research." In: *Elife* 3. https://doi.org/10.7554/eLife.04333.

Fisher, R A (1925). *Statistical Methods for Research Workers*. Edinburgh, Scotland: Oliver & Boyd.

Fisher, R A (1956). *Statistical Methods and Scientific Inference*. New York: Hafner.

Galak, Jeff et al. (2012). "Correcting the past: Failures to replicate PSI." In: *J Pers Soc Psychol* 103.6, pp. 933–48. https://doi.org/10.1037/a0029709.

Galton, Francis (1883). *Inquiries into Human Faculty and Its Development.* New York: Macmillan and Co.

Gerrig, R J and P G Zimbardo (2002). *Psychology and Life.* 16th ed. Boston: Allyn & Bacon.

Gigerenzer, Gerd (2004). "Mindless statistics." In: *J Socio-Econ* 33.5, pp. 587–606.

Godfrey-Smith, Peter (2021). *Theory and Reality: An Introduction to the Philosophy of Science.* 2nd ed. Chicago: University of Chicago Press.

Huberty, Lisa E, Katherine L Gross, and Carolyn J Miller (1998). "Effects of nitrogen addition on successional dynamics and species diversity in Michigan old-fields." In: *J Ecol* 86.5, 794–803. https://doi.org/10.1046/j.1365-2745.1998.8650794.x.

Ioannidis, John P A (2005). "Why most published research findings are false." In: *PLoS Med* 2.8, e124. https://doi.org/10.1371/journal.pmed.0020124.

Kaplan, Robert M and Veronica L Irvin (2015). "Likelihood of null effects of large NHLBI clinical trials has increased over time." In: *PLoS One* 10.8, e0132382. https://doi.org/10.1371/journal.pone.0132382.

Kass, Robert E and Adrian E Raftery (1995). "Bayes Factors." In: *J Am Stat Assoc* 90.430, pp. 773–95.

Kerr, N L (1998). "HARKing: Hypothesizing after the results are known." In: *Pers Soc Psychol Rev* 2.3, pp. 196–217. https://doi.org/10.1207/s15327957pspr0203_4.

Lee, Stephanie M (2018). "Here's how Cornell scientist Brian Wansink turned shoddy data into viral studies about how we eat." https://www.buzzfeednews.com/article/stephaniemlee/brian-wansink-cornell-p-hacking.

Lewandowsky, Stephan, Gilles E Gignac, and Klaus Oberauer (2013). "The role of conspiracist ideation and worldviews in predicting rejection of science." In: *PLoS One* 8.10, e75637. https://doi.org/10.1371/journal.pone.0075637.

Lewandowsky, Stephan, Gilles E Gignac, and Klaus Oberauer (2015). "Correction: The role of conspiracist ideation and worldviews in predicting rejection of science." In: *PLoS One* 10.8, e0134773. https://doi.org/10.1371/journal.pone.0134773.

Liebst, Lasse S et al. (2022). "Face-touching behaviour as a possible correlate of mask-wearing: A video observational study of public place incidents during the COVID-19 pandemic." In: *Transbound Emerg Dis* 69.3, pp. 1319–25. https://doi.org/10.1111/tbed.14094.

Luce, R D (1988). Review of *Cognition as Intuitive Statistics*, by G Gigerenzer and D J Murray. In: *Contemp Psychol* 33.7, pp. 582–83.

Majumder, M (2017). "Higher rates of hate crimes are tied to income inequality." https://fivethirtyeight.com/features/higher-rates-of-hate-crimes-are-tied-to-income-inequality/.

Marsh, Herbert W et al. (2014). "Exploratory structural equation modeling: An integration of the best features of exploratory and confirmatory factor analysis." In: *Annu Rev Clin Psychol* 10, 85–110. https://doi.org/10.1146/annurev-clinpsy-032813-153700.

McNeish, Daniel and Ken Kelley (2019). "Fixed effects models versus mixed effects models for clustered data: Reviewing the approaches, disentangling the differences, and making recommendations." In: *Psychol Methods* 24.1, pp. 20–35. https://doi.org/10.1037/met0000182.

Neyman, J (1937). "Outline of a theory of statistical estimation based on the classical theory of probability." In: *Phil Trans R Soc A* 236.767, pp. 333–80. https://doi.org/10.1098/rsta.1937.0005.

Neyman, J and E Pearson (1933). "On the problem of the most efficient tests of statistical hypotheses." In: *Phil Trans R Soc A* 231.694–706, pp. 289–337. https://doi.org/10.1098/rsta.1933.0009.

Open Science Collaboration (2015). "Psychology: Estimating the reproducibility of psychological science." In: *Science* 349.6251, aac4716. https://doi.org/10.1126/science.aac4716.

Pesch, Beate et al. (2012). "Cigarette smoking and lung cancer—relative risk estimates for the major histological types from a pooled analysis of case-control studies." In: *Int J Cancer* 131.5, 1210–19. https://doi.org/10.1002/ijc.27339.

Pew Research Center (2020). https://www.pewresearch.org/fact-tank/2020/11/20/facts-about-crime-in -the-u-s/.

Press, W H et al. (1992). *Numerical Recipes in C: The Art of Scientific Computing*. New York: Cambridge University Press.

Reinhart, Carmen M and Kenneth S Rogoff (2010). "Growth in a time of debt." In: *Am Econ Rev* 100.2, pp. 573–78. https://doi.org/10.1257/aer.100.2.573.

Schenker, Nathaniel and Jane F Gentleman (2013). "On Judging the significance of differences by examining the overlap between confidence intervals." In: *Am Statistician* 55.3, 182–186. http://www.jstor.org /stable/2685796.

Schoenfeld, Jonathan D and John P A Ioannidis (2013). "Is everything we eat associated with cancer? A systematic cookbook review." In: *Am J Clin Nutr* 97.1, 127–34. https://doi.org/10.3945/ajcn.112 .047142.

Simmons, Joseph P, Leif D Nelson, and Uri Simonsohn (2011). "False-positive psychology: Undisclosed flexibility in data collection and analysis allows presenting anything as significant." In: *Psychol Sci* 22.11, pp. 1359–66. https://doi.org/10.1177/0956797611417632.

Smaldino, Paul E and Richard McElreath (2016). "The natural selection of bad science." In: *R Soc Open Sci* 3.9, p. 160384. https://doi.org/10.1098/rsos.160384.

Stigler, Stephen M (2016). *The Seven Pillars of Statistical Wisdom*. Cambridge, MA: Harvard University Press.

Sullivan, Gail M and Richard Feinn (2012). "Using effect size—or why the P value is not enough." In: *J Grad Med Educ* 4.3, pp. 279–82. https://doi.org/10.4300/JGME-D-12-00156.1.

Teicholz, Nina (2014). *The Big Fat Surprise*. New York: Simon & Schuster.

Wagenmakers, Eric-Jan (2007). "A practical solution to the pervasive problems of p values." In: *Psychon Bull Rev* 14.5, pp. 779–804. https://doi.org/10.3758/bf03194105.

Wakefield, A J (1999). "MMR vaccination and autism." In: *Lancet* 354.9182, 949–50. https://doi.org /10.1016/S0140-6736(05)75696-8.

Wansink, Brian, David R Just, and Collin R Payne (2012). "Can branding improve school lunches?" In: *Arch Pediatr Adolesc Med* 166.10, pp. 1–2. https://doi.org/10.1001/archpediatrics.2012.999.

Woodfine, J D and D A Redelmeier (2015). "Berkson's paradox in medical care." In: *J Intern Med* 278.4, pp. 424–26. https://doi.org/10.1111/joim.12363.

INDEX

Milton Keynes UK
Ingram Content Group UK Ltd.
UKHW051222021223
433609UK00004B/13